Cram101 Textbook Outlines to accompany:

General Chemistry

Darrell Ebbing, Steven D. Gammon, 9th Edition

A Cram101 Inc. publication (c) 2011.

Cram101 Textbook Outlines and Cram101.com are Cram101 Inc. publications and services. All notes, highlights, reviews, and practice tests are written and prepared by Cram101, all rights reserved.

PRACTICE EXAMS.

Get all of the self-teaching practice exams for each chapter of this textbook at **www.Cram101.com** and ace the tests. Here is an example:

Chapter 1

General Chemistry
Darrell Ebbing, Steven D. Gammon, 9th Edition,
All Material Written and Prepared by Cram101

1. _____ is a chemical element with the symbol Mg, atomic number 12, atomic weight 24.3050 and common oxidation number +2.

   _____, an alkaline earth metal, is the ninth most abundant element in the universe by mass. The commonness of _____ is related to the fact that it is easily built up in supernova stars from a sequential addition of three helium nuclei to carbon.

   ○ Magnesium   ○ M.S. Factory, Valley
   ○ Maalox   ○ Machine drawn cylinder sheet

2. _____ is the most common metalloid. It is a chemical element, which has the symbol Si and atomic number 14. The atomic mass is 28.0855.

   ○ Silicon   ○ S process
   ○ S.O.S   ○ S_N1 reaction

3. A _____ in this article, is a physical model that represents molecules and their processes. The creation of mathematical models of molecular properties and behaviour is _____ ling, and their graphical depiction is molecular graphics, but these topics are closely linked and each uses techniques from the others. In this article, _____ will primarily refer to systems containing more than one atom and where nuclear structure is neglected.

   ○ Molecular model   ○ M.S. Factory, Valley

You get a 50% discount for the online exams. Go to **Cram101.com**, click Sign Up at the top of the screen, and enter DK73DW6022 in the promo code box on the registration screen. Access to Cram101.com is $4.95 per month, cancel at any time.

With Cram101.com online, you also have access to extensive reference material.

You will nail those essays and papers. Here is an example from a Cram101 Biology text:

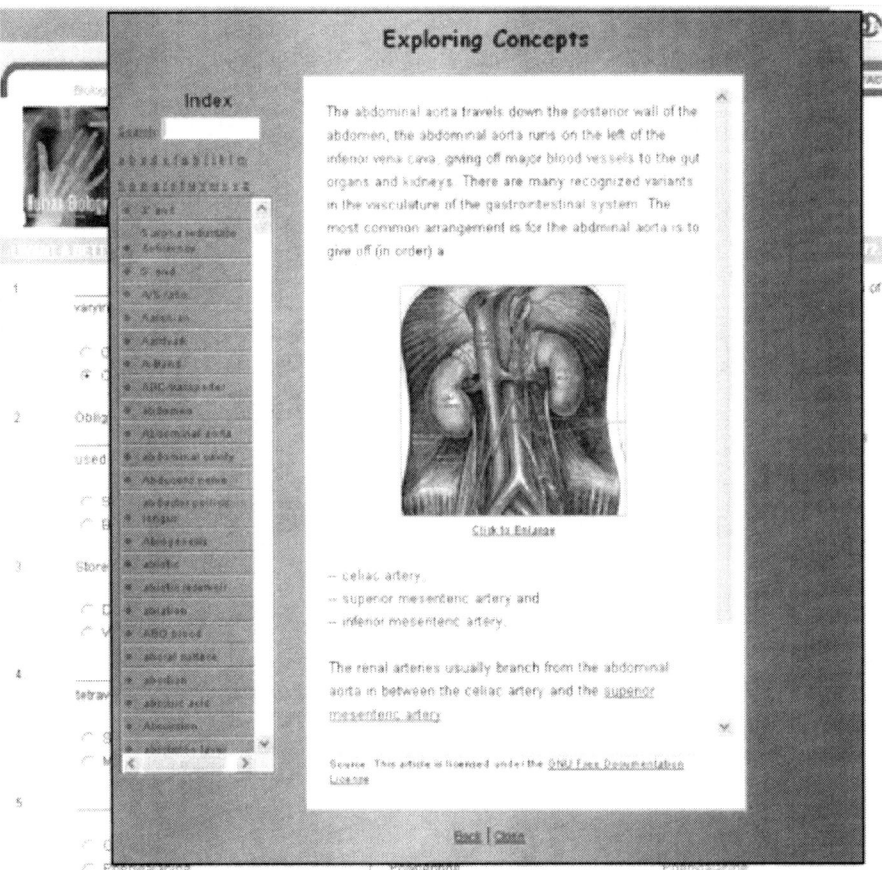

Visit **www.Cram101.com**, click Sign Up at the top of the screen, and enter DK73DW6022 in the promo code box on the registration screen. Access to www.Cram101.com is normally $9.95 per month, but because you have purchased this book, your access fee is only $4.95 per month, cancel at any time. Sign up and stop highlighting textbooks forever.

Learning System

Cram101 Textbook Outlines is a learning system. The notes in this book are the highlights of your textbook, you will never have to highlight a book again.

How to use this book. Take this book to class, it is your notebook for the lecture. The notes and highlights on the left hand side of the pages follow the outline and order of the textbook. All you have to do is follow along while your instructor presents the lecture. Circle the items emphasized in class and add other important information on the right side. With Cram101 Textbook Outlines you'll spend less time writing and more time listening. Learning becomes more efficient.

Cram101.com Online

Increase your studying efficiency by using Cram101.com's practice tests and online reference material. It is the perfect complement to Cram101 Textbook Outlines. Use self-teaching matching tests or simulate in-class testing with comprehensive multiple choice tests, or simply use Cram's true and false tests for quick review. Cram101.com even allows you to enter your in-class notes for an integrated studying format combining the textbook notes with your class notes.

Visit **www.Cram101.com**, click Sign Up at the top of the screen, and enter **DK73DW6022** in the promo code box on the registration screen. Access to www.Cram101.com is normally $9.95 per month, but because you have purchased this book, your access fee is only $4.95 per month. Sign up and stop highlighting textbooks forever.

Copyright © 2011 by Cram101, Inc. All rights reserved. "Cram101"® and "Never Highlight a Book Again!"® are registered trademarks of Cram101, Inc. ISBN(s): 9781616986025. EDR-5.2010111

General Chemistry
Darrell Ebbing, Steven D. Gammon, 9th

CONTENTS

1. Chemistry and Measurement 2
2. Atoms, Molecules, and Ions 14
3. Calculations with Chemical Formulas and Equations 50
4. Chemical Reactions 64
5. The Gaseous State 88
6. Thermochemistry 104
7. Quantum Theory of the Atom 120
8. Electron Configurations and Periodicity 132
9. Ionic and Covalent Bonding 162
10. Molecular Geometry and Chemical Bonding Theory 186
11. States of Matter; Liquids and Solids 206
12. Solutions 230
13. Rates of Reaction 248
14. Chemical Equilibrium 268
15. Acids and Bases 280
16. Acid—Base Equilibria 296
17. Solubility and Complex-Ion Equilibria 308
18. Thermodynamics and Equilibrium 320
19. Electrochemistry 328
20. Nuclear Chemistry 342
21. Chemistry of the Main-Group Elements 364
22. The Transition Elements and Coordination Compounds 412
23. Organic Chemistry 432
24. Polymer Materials: Synthetic and Biological 444

Chapter 1. Chemistry and Measurement

Magnesium	Magnesium is a chemical element with the symbol Mg, atomic number 12, atomic weight 24.3050 and common oxidation number +2. Magnesium, an alkaline earth metal, is the ninth most abundant element in the universe by mass. The commonness of Magnesium is related to the fact that it is easily built up in supernova stars from a sequential addition of three helium nuclei to carbon .
Silicon	Silicon is the most common metalloid. It is a chemical element, which has the symbol Si and atomic number 14. The atomic mass is 28.0855.
Molecular model	A Molecular model in this article, is a physical model that represents molecules and their processes. The creation of mathematical models of molecular properties and behaviour is Molecular model ling, and their graphical depiction is molecular graphics, but these topics are closely linked and each uses techniques from the others. In this article, Molecular model will primarily refer to systems containing more than one atom and where nuclear structure is neglected.
Oxygen	Oxygen and -γενÎ®ς (-genÄ"s) (producer, literally begetter) is the element with atomic number 8 and represented by the symbol O. It is a member of the chalcogen group on the periodic table, and is a highly reactive nonmetallic period 2 element that readily forms compounds (notably oxides) with almost all other elements. At standard temperature and pressure two atoms of the element bind to form di Oxygen , a colorless, odorless, tasteless diatomic gas with the formula O_2. Oxygen is the third most abundant element in the universe by mass after hydrogen and helium and the most abundant element by mass in the Earth"s crust.
Chemistry	In the history of science, the etymology of the word Chemistry is a debatable issue. It is agreed that the word "alchemy" is a European one, derived from Arabic, but the origin of the root word, chem, is uncertain. Words similar to it have been found in most ancient languages, with different meanings, but conceivably somehow related to alchemy.
Atomic weight	Atomic weight (symbol: A_r) is a dimensionless physical quantity, the ratio of the average mass of atoms of an element to 1/12 of the mass of an atom of carbon-12. The term is usually used, without further qualification, to refer to the standard Atomic weight s published at regular intervals by the International Union of Pure and Applied Chemistry (IUPAC) and which are intended to be applicable to normal laboratory materials
Molecule	A Molecule is defined as a sufficiently stable, electrically neutral group of at least two atoms in a definite arrangement held together by very strong (covalent) chemical bonds. Molecule s are distinguished from polyatomic ions in this strict sense. In organic chemistry and biochemistry, the term Molecule is used less strictly and also is applied to charged organic Molecule s and bio Molecule s.
Gallium	Gallium is a chemical element that has the symbol Ga and atomic number 31. Elemental Gallium does not occur in nature, but as the Ga salt, in trace amounts in bauxite and zinc ores. A soft silvery metallic poor metal, elemental Gallium is a brittle solid at low temperatures.

Chapter 1. Chemistry and Measurement

Chapter 1. Chemistry and Measurement

Gallium arsenide	Gallium arsenide is a compound of two elements, gallium and arsenic. It is an important semiconductor and is used to make devices such as microwave frequency integrated circuits (ie, MMICs), infrared light-emitting diodes, laser diodes and solar cells. Gallium arsenide can be prepared from the elements and a number of industrial processes use this, for example: - the crystal growth using a horizontal zone furnace (Bridgman-Stockbarger technique) where Ga and Arsenic vapor react and deposit on a seed crystal at the cooler end of the furnace. - LEC (liquid encapsulated Czochralski) growth Alternative methods for producing films of GaAs include: - VPE reaction of gaseous gallium metal and arsenic trichloride $2Ga + 2AsCl_3 \rightarrow 2GaAs + 3Cl_2$ - MOCVD reaction of trimethylgallium and arsine: $Ga(CH_3)_3 + AsH_3 \rightarrow GaAs + 3CH_4$ Wet etching of GaAs industrially uses an oxidizing agent e.g. hydrogen peroxide or bromine water, and the same strategy has been described in a patent relating to processing scrap components containing GaAs where the Ga^{3+} is complexed with a hydroxamic acid, "HA" e.g.: $GaAs + H_2O_2 + \text{"HA"} \rightarrow \text{"GaA"} \text{ complex} + H_3AsO_4 + 4H_2O$ Oxidation of GaAs occurs in air and degrades performance of the semiconductor, the surface can be passivated by depositing a cubic gallium(II) sulfide layer using a tert-butyl gallium sulfide compound such as $(^tBuGaS)_7$. GaAs has some electronic properties which are superior to those of silicon.
Acid	An Acid is traditionally considered any chemical compound that, when dissolved in water, gives a solution with a hydrogen ion activity greater than in pure water, i.e. a pH less than 7.0. That approximates the modern definition of Johannes Nicolaus Brønsted and Martin Lowry, who independently defined an Acid as a compound which donates a hydrogen ion (H^+) to another compound (called a base.) Common examples include acetic Acid and sulfuric Acid (used in car batteries.)

Chapter 1. Chemistry and Measurement

Chapter 1. Chemistry and Measurement

Atom	The Atom is a basic unit of matter consisting of a dense, central nucleus surrounded by a cloud of negatively charged electrons. The atomic nucleus contains a mix of positively charged protons and electrically neutral neutrons (except in the case of hydrogen-1, which is the only stable nuclide with no neutron.) The electrons of an Atom are bound to the nucleus by the electromagnetic force.
Silver	Silver is a chemical element with the chemical symbol Ag and atomic number 47. A soft, white, lustrous transition metal, it has the highest electrical conductivity of any element and the highest thermal conductivity of any metal. The metal occurs naturally in its pure, free form , as an alloy with gold (electrum) and other metals, and in minerals such as argentite and chlorargyrite.
Combustion	Combustion or burning is a complex sequence of exothermic chemical reactions between a fuel (usually a hydrocarbon) and an oxidant accompanied by the production of heat or both heat and light in the form of either a glow or flames, appearance of light flickering. Direct Combustion by atmospheric oxygen is a reaction mediated by radical intermediates. The conditions for radical production are naturally produced by thermal runaway, where the heat generated by Combustion is necessary to maintain the high temperature necessary for radical production.
Mercury	There are seven isotopes of mercury with Hg-202 being the most abundant (29.86%.) The longest-lived radioisotopes are ^{194}Hg with a half-life of 444 years, and ^{203}Hg with a half-life of 46.612 days. Most of the remaining radioisotopes have half-lives that are less than a day.
Mercury oxide	Mercury oxide has a formula of HgO and a formula weight of 216.6. It has a red or orange color. Mercury oxide is a solid at room temperature and pressure.
Acid-base reaction	An Acid-base reaction is a chemical reaction that occurs between an acid and a base. Several concepts that provide alternative definitions for the reaction mechanisms involved and their application in solving related problems exist. Despite several differences in definitions, their importance becomes apparent as different methods of analysis when applied to Acid-base reaction s for gaseous or liquid species, or when acid or base character may be somewhat less apparent.
Chemical bond	A Chemical bond is the physical process responsible for the attractive interactions between atoms and molecules, and that which confers stability to diatomic and polyatomic chemical compounds. The explanation of the attractive forces is a complex area that is described by the laws of quantum electrodynamics. In practice, however, chemists usually rely on quantum theory or qualitative descriptions that are less rigorous but more easily explained to describe Chemical bond ing.

Chapter 1. Chemistry and Measurement

Chapter 1. Chemistry and Measurement

Chemical change	In a Chemical change bonds are broken and new bonds are formed between different atoms. This breaking and forming of bonds takes place when particles of the original materials collide with one another. Some exothermic reactions may be hot enough to cause certain chemicals to also undergo a change in state; for example in the case of aqueous solutions, bubbles may not necessarily be newly produced gas but instead water vapor.
Impurities	Impurities are substances inside a confined amount of liquid, gas which differ from the chemical composition of the material or compound. Impurities are either naturally occurring or added during synthesis of a chemical or commercial product. During production, Impurities may be purposely, accidentally, inevitably, or incidentally added into the substance.
Water	Water is the most abundant molecule on Earth"s surface, constituting about 75% of the Earth"s surface. In nature it exists in liquid, solid, and gaseous states. It is in dynamic equilibrium between the liquid and gas states at standard temperature and pressure.
Intermolecular forces	In physics, chemistry, and biology, Intermolecular forces are forces that act between stable molecules or between functional groups of macromolecules. Intermolecular forces include momentary attractions between molecules, diatomic free elements, and individual atoms. They differ from covalent and ionic bonding in that they are not stable, but are caused by momentary polarization of particles.
Distillation	Distillation is a method of separating mixtures based on differences in their volatilities in a boiling liquid mixture. Distillation is a unit operation, or a physical separation process, and not a chemical reaction. Commercially, Distillation has a number of uses.
Iron	Iron is a chemical element with the symbol Fe and atomic number 26. Iron is a group 8 and period 4 element. Iron and Iron alloys (steels) are by far the most common metals and the most common ferromagnetic materials in everyday use.
Sodium	Sodium is a metallic element with a symbol Na and atomic number 11. It is a soft, silvery-white, highly reactive metal and is a member of the alkali metals within "group 1" (formerly known as "group IA".) It has only one stable isotope, ^{23}Na.
Vapor	A Vapor or vapour is a substance in the gas phase at a temperature lower than its critical temperature. This means that the Vapor can be condensed to a liquid or to a solid by increasing its pressure, without reducing the temperature. For example, water has a critical temperature of 374°C (or 647 K) which is the highest temperature at which liquid water can exist.
Potassium	Potassium is a chemical element. It has the symbol K, atomic number 19, and atomic mass 39.0983. Potassium was first isolated from potash.

Chapter 1. Chemistry and Measurement

Chapter 1. Chemistry and Measurement

Law of definite proportions	In chemistry, the Law of definite proportions and also the elements states that a chemical compound always contains exactly the same proportion of elements by mass. An equivalent statement is the law of constant composition, which states that all samples of a given chemical compound have the same elemental composition. For example, oxygen makes up $8/9$ of the mass of any sample of pure water, while hydrogen makes up the remaining $1/9$ of the mass.
Solution	In chemistry, a Solution is a homogeneous mixture composed of two or more substances. In such a mixture, a solute is dissolved in another substance, known as a solvent. Gases may dissolve in liquids, for example, carbon dioxide or oxygen in water.
Activity	In chemical thermodynamics Activity is a measure of the "effective concentration" of a species in a mixture. By convention, it is a dimensionless quantity. The Activity of pure substances in condensed phases (solid or liquids) is normally taken as unity.
Ion	An Ion is an atom or molecule where the total number of electrons is not equal to the total number of protons, giving it a net positive or negative electrical charge. Since protons are positively charged and electrons are negatively charged, if there are more electrons than protons, the atom or molecule will be negatively charged. This is called an an Ion , from the Greek á¼€vÎ¬ , meaning "up".
Chromatography	Chromatography is the collective term for a set of laboratory techniques for the separation of mixtures. It involves passing a mixture dissolved in a "mobile phase" through a stationary phase, which separates the analyte to be measured from other molecules in the mixture and allows it to be isolated. Chromatography may be preparative or analytical.
Column chromatography	Column chromatography in chemistry is a method used to purify individual chemical compounds from mixtures of compounds. It is often used for preparative applications on scales from micrograms up to kilograms. The classical preparative chromatography column is a glass tube with a diameter from 50 mm and a height of 50 cm to 1 m with a tap at the bottom.
Ion-exchange chromatography	Ion-exchange chromatography is a process that allows the separation of ions and polar molecules based on the charge properties of the molecules. It can be used for almost any kind of charged molecule including large proteins, small nucleotides and amino acids. The solution to be injected is usually called a sample, and the individually separated components are called analytes.
Thymol blue	Thymol blue is a brownish-green or reddish-brown crystaline powder that is used as an pH indicator. It is insoluble in water but soluble in alcohol and dilute alkali solutions. It transitions from red to yellow at pH 1.2-2.8 and from yellow to blue from at pH 8.0-9.6.

Chapter 1. Chemistry and Measurement

Chapter 1. Chemistry and Measurement

Base	In chemistry, a Base is most commonly thought of as an aqueous substance that can accept hydrogen ions. A Base is also often referred to as an alkali if OH^- ions are involved. This refers to the Brønsted-Lowry theory of acids and bases.
Concentration	In chemistry, Concentration is the measure of how much of a given substance there is mixed with another substance. This can apply to any sort of chemical mixture, but most frequently the concept is limited to homogeneous solutions, where it refers to the amount of solute in the solvent. To concentrate a solution, one must add more solute, or reduce the amount of solvent (for instance, by selective evaporation.)
Molar concentration	In chemistry, molar concentration is a measure of the concentration of a solute in a solution ionic in thermodynamics the use of molar concentration is often not very convenient, because the volume of most solutions slightly depends on temperature due to thermal expansion. This problem is usually resolved by introducing temperature correction factors, or by using a temperature-independent measure of concentration such as molality.
Molecular mass	The Molecular mass of a substance, frequently referred by the older term molecular weight and abbreviated as MW, is the mass of one molecule of that substance, relative to the unified atomic mass unit u (equal to 1/12 the mass of one isotope of carbon-12.) This is distinct from the relative Molecular mass of a molecule, which is the ratio of the mass of that molecule to 1/12 of the mass of carbon 12 and is a dimensionless number. Relative Molecular mass is abbreviated to M_r.
Toluene	Toluene phenylmethane, and Toluol, is a clear water-insoluble liquid with the typical smell of paint thinners, redolent of the sweet smell of the related compound benzene. It is an aromatic hydrocarbon that is widely used as an industrial feedstock and as a solvent. Like other solvents, Toluene is also used as an inhalant drug for its intoxicating properties; however this causes severe neurological harm.
Mole	The mole is a unit of amount of substance: it is an SI base unit, and one of the few units used to measure this physical quantity. The name "mole" was coined in German by Wilhelm Ostwald in 1893, although the related concept of equivalent mass had been in use at least a century earlier. The name is assumed to be derived from the word Molekül (molecule.)
Moles	Moles are the majority of the members of the mammal family Talpidae in the order Soricomorpha. Although most moles burrow, some species are aquatic or semi-aquatic. moles have cylindrical bodies covered in fur, with small or covered eyes; the ears are generally not visible.

Chapter 2. Atoms, Molecules, and Ions

Atomic theory	In chemistry and physics, Atomic theory is a theory of the nature of matter, which states that matter is composed of discrete units called atoms, as opposed to the obsolete notion that matter could be divided into any arbitrarily small quantity. It began as a philosophical concept in ancient Greece and India and entered the scientific mainstream in the early 19th century when discoveries in the field of chemistry showed that matter did indeed behave as if it were made up of particles. The word "atom" was applied to the basic particle that constituted a chemical element, because the chemists of the era believed that these were the fundamental particles of matter.
Dalton	Dalton is an ab initio quantum chemistry software program. It is capable of calculating various molecular properties using the Hartree-Fock, MP2, MCSCF and coupled cluster theories. Version 2.0 of Dalton added support for density functional theory calculations.
Sodium	Sodium is a metallic element with a symbol Na and atomic number 11. It is a soft, silvery-white, highly reactive metal and is a member of the alkali metals within "group 1" (formerly known as "group IA".) It has only one stable isotope, ^{23}Na.
Chlorin	In organic chemistry, a Chlorin is a large heterocyclic aromatic ring consisting, at the core, of three pyrroles and one pyrroline coupled through four methine linkages. Unlike a porphyrin, a Chlorin is therefore largely aromatic but not aromatic through the entire circumference of the ring. Magnesium-containing Chlorin s are called chlorophylls, and are the central photosensitive pigment in chloroplasts.
Chlorine	Chlorine . As the chloride ion, which is part of common salt and other compounds, it is abundant in nature and necessary to most forms of life, including humans. In its elemental form (Cl_2 or "di Chlorine ") under standard conditions, Chlorine is a powerful oxidant and is used in bleaching and disinfectants.
Atomic weight	Atomic weight (symbol: A_r) is a dimensionless physical quantity, the ratio of the average mass of atoms of an element to 1/12 of the mass of an atom of carbon-12. The term is usually used, without further qualification, to refer to the standard Atomic weight s published at regular intervals by the International Union of Pure and Applied Chemistry (IUPAC) and which are intended to be applicable to normal laboratory materials
Molecular model	A Molecular model in this article, is a physical model that represents molecules and their processes. The creation of mathematical models of molecular properties and behaviour is Molecular model ling, and their graphical depiction is molecular graphics, but these topics are closely linked and each uses techniques from the others. In this article, Molecular model will primarily refer to systems containing more than one atom and where nuclear structure is neglected.

Chapter 2. Atoms, Molecules, and Ions

Chapter 2. Atoms, Molecules, and Ions

Water	Water is the most abundant molecule on Earth"s surface, constituting about 75% of the Earth"s surface. In nature it exists in liquid, solid, and gaseous states. It is in dynamic equilibrium between the liquid and gas states at standard temperature and pressure.
Acid-base reaction	An Acid-base reaction is a chemical reaction that occurs between an acid and a base. Several concepts that provide alternative definitions for the reaction mechanisms involved and their application in solving related problems exist. Despite several differences in definitions, their importance becomes apparent as different methods of analysis when applied to Acid-base reaction s for gaseous or liquid species, or when acid or base character may be somewhat less apparent.
Activity	In chemical thermodynamics Activity is a measure of the "effective concentration" of a species in a mixture. By convention, it is a dimensionless quantity. The Activity of pure substances in condensed phases (solid or liquids) is normally taken as unity.
Ion	An Ion is an atom or molecule where the total number of electrons is not equal to the total number of protons, giving it a net positive or negative electrical charge. Since protons are positively charged and electrons are negatively charged, if there are more electrons than protons, the atom or molecule will be negatively charged. This is called an an Ion , from the Greek á¼€vÎ¬ , meaning "up".
Carbon	Carbon is the chemical element with symbol C and atomic number 6. As a member of group 14 on the periodic table, it is nonmetallic and tetravalent--making four electrons available to form covalent chemical bonds. There are three naturally occurring isotopes, with ^{12}C and ^{13}C being stable, while ^{14}C is radioactive, decaying with a half-life of about 5730 years.
Electron	The Electron is a subatomic particle that carries a negative electric charge. It has no known substructure and is believed to be a point particle. An Electron has a mass that is approximately 1836 times less than that of the proton.
Law of definite proportions	In chemistry, the Law of definite proportions and also the elements states that a chemical compound always contains exactly the same proportion of elements by mass. An equivalent statement is the law of constant composition, which states that all samples of a given chemical compound have the same elemental composition. For example, oxygen makes up $^8/_9$ of the mass of any sample of pure water, while hydrogen makes up the remaining $^1/_9$ of the mass.
Law of multiple proportions	The Law of multiple proportions is one of the basic laws in chemistry, and is a major tool of chemical measurement (stoichiometry.) This law states that when elements combine they do so in a ratio of small whole numbers. For example, carbon and oxygen react to form CO or CO_2, but not $CO_{1.3}$ for instance.

Chapter 2. Atoms, Molecules, and Ions

Chapter 2. Atoms, Molecules, and Ions

Oxygen	Oxygen and -γενî®ς (-genÄ"s) (producer, literally begetter) is the element with atomic number 8 and represented by the symbol O. It is a member of the chalcogen group on the periodic table, and is a highly reactive nonmetallic period 2 element that readily forms compounds (notably oxides) with almost all other elements. At standard temperature and pressure two atoms of the element bind to form di Oxygen , a colorless, odorless, tasteless diatomic gas with the formula O_2. Oxygen is the third most abundant element in the universe by mass after hydrogen and helium and the most abundant element by mass in the Earth"s crust.
Acid	An Acid is traditionally considered any chemical compound that, when dissolved in water, gives a solution with a hydrogen ion activity greater than in pure water, i.e. a pH less than 7.0. That approximates the modern definition of Johannes Nicolaus Brønsted and Martin Lowry, who independently defined an Acid as a compound which donates a hydrogen ion (H^+) to another compound (called a base.) Common examples include acetic Acid and sulfuric Acid (used in car batteries.)
Atom	The Atom is a basic unit of matter consisting of a dense, central nucleus surrounded by a cloud of negatively charged electrons. The atomic nucleus contains a mix of positively charged protons and electrically neutral neutrons (except in the case of hydrogen-1, which is the only stable nuclide with no neutron.) The electrons of an Atom are bound to the nucleus by the electromagnetic force.
Atomic orbital	An Atomic orbital is a mathematical function that describes the wave-like behavior of either one electron or a pair of electrons, in an atom. This function can be used to calculate the probability of finding any electron of an atom in any specific region around the atom"s nucleus. These functions may serve as three-dimensional graph of an electron"s likely location.
Molecule	A Molecule is defined as a sufficiently stable, electrically neutral group of at least two atoms in a definite arrangement held together by very strong (covalent) chemical bonds. Molecule s are distinguished from polyatomic ions in this strict sense. In organic chemistry and biochemistry, the term Molecule is used less strictly and also is applied to charged organic Molecule s and bio Molecule s.
Nuclear model	A Nuclear model is any model that attempts to describe the atomic nucleus.

Chapter 2. Atoms, Molecules, and Ions

Chapter 2. Atoms, Molecules, and Ions

List of known Nuclear model s:

- Alpha particle model
- Cluster model
- Droplet model
- Fermi gas model
- Independent Particle model
- Interacting boson model
- Lattice Gas Model (FCC ' SCP)
- Liquid drop model
- Nuclear Lattice model
- Quark model
- Shell model
- Variable Phase model

Atom.

Oil-drop experiment

In 1909, Robert Millikan and Harvey Fletcher performed the Oil-drop experiment to measure the elementary electric charge (the charge of the electron.) The experiment entailed balancing the downward gravitational force with the upward buoyant and electric forces on tiny charged droplets of oil suspended between two metal electrodes. Since the density of the oil was known, the droplets" masses, and therefore their gravitational and buoyant forces, could be determined from their observed radii.

Rutherford

The Rutherford is an obsolete unit of radioactivity, defined as the activity of a quantity of radioactive material in which one million nuclei decay per second. It is therefore equivalent to one megabecquerel. It was named after Ernest Rutherford It is not an SI unit.

Emission

In physics, Emission is the process by which the energy of a photon is released by another entity, for example, by an atom whose electrons make a transition between two electronic energy levels. The emitted energy is in the form of a photon.

The emittance of an object quantifies how much light is emitted by it.

Atomic number

In chemistry and physics, the Atomic number is the number of protons found in the nucleus of an atom and therefore identical to the charge number of the nucleus. It is conventionally represented by the symbol Z. The Atomic number uniquely identifies a chemical element. In an atom of neutral charge, Atomic number is equal to the number of electrons.

Hydrogen

Hydrogen is the chemical element with atomic number 1. It is represented by the symbol H. At standard temperature and pressure, Hydrogen is a colorless, odorless, nonmetallic, tasteless, highly flammable diatomic gas with the molecular formula H_2. With an atomic weight of 1.007 94 u, Hydrogen is the lightest element.

Chapter 2. Atoms, Molecules, and Ions

Chapter 2. Atoms, Molecules, and Ions

Nuclear reaction	In nuclear physics, a Nuclear reaction is the process in which two nuclei or nuclear particles collide to produce products different from the initial particles. In principle a reaction can involve more than three particles colliding, but because the probability of three or more nuclei to meet at the same time at the same place is much less than for two nuclei, such an event is exceptionally rare. While the transformation is spontaneous in the case of radioactive decay, it is initiated by a particle in the case of a Nuclear reaction.
Isotopes	Isotopes are any of the different types of atoms of the same chemical element, each having a different atomic mass (mass number.) Isotopes of an element have nuclei with the same number of protons (the same atomic number) but different numbers of neutrons. Therefore, Isotopes of the same element have different mass numbers (number of nucleons.)
Mass number	The Mass number is the total number of protons and neutrons in an atomic nucleus. Because protons and neutrons both are baryons, the Mass number A is identical with the baryon number B as of the nucleus as of the whole atom or ion. The Mass number is different for each different isotope of a chemical element.
Metallurgy	Metallurgy is a domain of materials science that studies the physical and chemical behavior of metallic elements, their intermetallic compounds, and their mixtures, which are called alloys. It is also the technology of metals: the way in which science is applied to their practical use. Metallurgy is commonly used in the craft of metalworking.
Nuclide	A Nuclide is an atomic nucleus characterized by its specific constitution, i.e., by its number of protons, its number of neutrons, and its energy content. The various Nuclide s, or species, of a particular chemical element with equal proton number , but different neutron numbers were called isotopes of the element, before the more inclusive term Nuclide was internationally accepted (ca. 1950.)
Ionization	Ionization is the physical process of converting an atom or molecule into an ion by adding or removing charged particles such as electrons or other ions. This is often confused with dissociation (chemistry.) The process works slightly differently depending on whether an ion with a positive or a negative electric charge is being produced.
Isotopes of Oxygen	There are three stable Isotopes of oxygen that lead to oxygen (O) having a standard atomic mass of 15.9994(3) u. Also 10 unstable isotopes have been characterized. Naturally occurring oxygen is composed of three stable isotopes, ^{16}O, ^{17}O, and ^{18}O, with ^{16}O being the most abundant (99.762% natural abundance.)

Chapter 2. Atoms, Molecules, and Ions

Chapter 2. Atoms, Molecules, and Ions

Atomic mass	The Atomic mass is the mass of an atom, most often expressed in unified Atomic mass units. The Atomic mass may be considered to be the total mass of protons, neutrons and electrons in a single atom (when the atom is motionless.) The Atomic mass is sometimes incorrectly used as a synonym of relative Atomic mass, average Atomic mass and atomic weight; however, these differ subtly from the Atomic mass.
Atomic mass unit	The unified Atomic mass unit sometimes, universal mass unit, is a unit of mass used to express atomic and molecular masses. It is the approximate mass of a hydrogen atom, a proton, or a neutron. The precise definition is that it is one twelfth of the mass of an isolated atom of carbon-12 (^{12}C) at rest and in its ground state.
Relative Atomic mass	The atomic mass (m_a) is the mass of an atom, most often expressed in unified atomic mass units. The atomic mass may be considered to be the total mass of protons, neutrons and electrons in a single atom (when the atom is motionless.) The atomic mass is sometimes incorrectly used as a synonym of Relative atomic mass, average atomic mass and atomic weight; however, these differ subtly from the atomic mass.
Neon	Neon is the chemical element that has the symbol Ne and atomic number 10. Although a very common element in the universe, it is rare on Earth. A colorless, inert noble gas under standard conditions, Neon gives a distinct reddish-orange glow when used in discharge tubes and Neon lamps.
Abundance	The abundance of a chemical element measures how relatively common the element is, or how much of the element there is by comparison to all other elements. abundance may be variously measured by the mass-fraction (the same as weight fraction), or mole-fraction (fraction of atoms, or sometimes fraction of molecules, in gases), or by volume fraction. Measurement by volume-fraction is a common abundance measure in mixed gases such as atmospheres, which is close to molecular mole-fraction for ideal gas mixtures (i.e., gas mixtures at relatively low densities and pressures.)
Mass spectrum	A Mass spectrum is an intensity vs. m/z (mass-to-charge ratio) plot representing a chemical analysis. Hence, the Mass spectrum of a sample is a pattern representing the distribution of components (atoms or molecules) by mass (more correctly: mass-to-charge ratio) in a sample. It is usually acquired using an instrument called a mass spectrometer.
Chromium	Chromium is a chemical element which has the symbol Cr and atomic number 24. It is a steely-gray, lustrous, hard metal that takes a high polish and has a high melting point. It is also odourless, tasteless, and malleable.

Chapter 2. Atoms, Molecules, and Ions

Chapter 2. Atoms, Molecules, and Ions

Periodic table	The periodiÑ table of the chemical elements (also, Periodic table of the elements or just Periodic table) is a tabular display of the chemical elements. Although precursors to this table exist, its invention is generally credited to Russian chemist Dmitri Mendeleev in 1869, who intended the table to illustrate recurring trends in the properties of the elements. The layout of the table has been refined and extended over time, as new elements have been discovered, and new theoretical models have been developed to explain chemical behavior.
Group	In chemistry, a Group is a vertical column in the periodic table of the chemical elements. The name family is derived from the fact that the elements share similar characteristics and traits, just as members of any human family would. There are 18 groups in the standard periodic table.
Alkali	In chemistry, an Alkali is a basic, ionic salt of an Alkali metal or Alkali ne earth metal element. Alkali s are best known for being bases that dissolve in water. Bases are compounds with a pH greater than 7.
Bromine	Bromine , Greek: βρά¿¶μος, brómos, meaning "stench "), is a chemical element with the symbol Br and atomic number 35. A halogen element, Bromine is a reddish-brown volatile liquid at standard room temperature that is intermediate in reactivity between chlorine and iodine. Bromine vapours are corrosive and toxic.
Germanium	Germanium is a chemical element with the symbol Ge and atomic number 32. It is a lustrous, hard, grayish-white metalloid in the carbon group, chemically similar to its group neighbors tin and silicon. Germanium has five naturally occurring isotopes ranging in atomic mass number from 70 to 76.
Halogen	The Halogen s or Halogen elements are a series of nonmetal elements from Group 17 IUPAC Style (formerly: VII, VIIA) of the periodic table, comprising fluorine, (F); chlorine, (Cl); bromine, (Br); iodine, (I); and astatine, (At.) The undiscovered element 117, temporarily named ununseptium, may also be a Halogen The group of Halogen s is the only periodic table group which contains elements in all three familiar states of matter at standard temperature and pressure.
Metal	In chemistry, a Metal is an element, compound, or alloy characterized by high electrical conductivity. In a Metal atoms readily lose electrons to form positive ions ; those ions are surrounded by delocalized electrons, which are responsible for the conductivity. The thus produced solid is held by electrostatic interactions between the ions and the electron cloud, which are called Metal lic bonds.
Metalloid	Metalloid or semi metal is a term used in chemistry when classifying the chemical elements. On the basis of their general physical and chemical properties, nearly every element in the periodic table can be termed either a metal or a nonmetal. However, a few elements with intermediate properties are referred to as Metalloid s .

Chapter 2. Atoms, Molecules, and Ions

Chapter 2. Atoms, Molecules, and Ions

Nonmetal	Nonmetal is a term used in chemistry when classifying the chemical elements. On the basis of their general physical and chemical properties, every element in the periodic table can be termed either a metal or a Nonmetal. (A few elements with intermediate properties are referred to as metalloids.)
Silicon	Silicon is the most common metalloid. It is a chemical element, which has the symbol Si and atomic number 14. The atomic mass is 28.0855.
Isomer	In chemistry, Isomer s are compounds with the same molecular formula but different structural formula. Isomer s do not necessarily share similar properties unless they also have the same functional groups. This should not be confused with a nuclear Isomer which involves a nucleus at different states of excitement.
Oxoacid	An Oxoacid is an acid which contains oxygen. More specifically, it is an acid which: 1. contains oxygen; 2. contains at least one other element; 3. has at least one hydrogen atom bound to oxygen; and 4. forms an ion by the loss of one or more protons. The name oxyacid is sometimes used, although this is not recommended. Generally, Oxoacid s are simply polyatomic ions with a hydrogen cation. Under Lavoisier"s original theory, all acids contained oxygen, which was named from the Greek οξυς (acid, sharp) and γεινομαι (geinomai) (engender.)
Chemical formula	A Chemical formula is a way of expressing information about the atoms that constitute a particular chemical compound, and how the relationship between those atoms changes in chemical reactions. For molecular compounds it is also known as the molecular formula, and identifies each constituent element by its chemical symbol and indicates the number of atoms of each element found in each discrete molecule of that compound. If a molecule contains more than one atom of a particular element, this quantity is indicated using a subscript after the chemical symbol (although 19th-century books often used superscripts.)
Radioactive decay	Radioactive decay is the process in which an unstable atomic nucleus spontaneously loses energy by emitting ionizing particles and radiation. This decay, or loss of energy, results in an atom of one type, called the parent nuclide transforming to an atom of a different type, called the daughter nuclide. For example: a carbon-14 atom (the "parent") emits radiation and transforms to a nitrogen-14 atom (the "daughter".)
Transuranium elements	In chemistry, Transuranium elements are the chemical elements with atomic numbers greater than 92 None of these elements are stable; they decay radioactively into other elements.

Chapter 2. Atoms, Molecules, and Ions

Chapter 2. Atoms, Molecules, and Ions

	Of the elements with atomic numbers 1 to 92, all but four occur in easily detectable quantities on earth, having stable, or very long half life isotopes, or are created as common products of the decay of uranium.
Ununbium	Ununbium is a chemical element in the periodic table that has the temporary symbol Uub and the atomic number 112. "Ununbium" is a IUPAC systematic element name, used until the element gets an accepted name.
	Ununbium was first created by the GSI in 1996, who propose the permanent name copernicium (Â·)) and the element symbol "Cp" for it.
Ammonia	Ammonia is a compound of nitrogen and hydrogen with the formula NH_3. It is normally encountered as a gas with a characteristic pungent odor. Ammonia contributes significantly to the nutritional needs of terrestrial organisms by serving as a precursor to foodstuffs and fertilizers.
Ununhexium	Ununhexium is the temporary name of a synthetic superheavy element in the periodic table that has the temporary symbol Uuh and has the atomic number 116.
	About 30 atoms of Ununhexium have been produced to date, either directly or as a decay product of ununoctium, and are associated with decays from the four neighbouring isotopes with masses 290-293. The most stable isotope to date is Uuh-293 with a half-life of 63 ms.
Ununquadium	Ununquadium is the temporary name of a radioactive chemical element in the periodic table that has the temporary symbol Uuq and has the atomic number 114.

Problems listening to this file? See media help.

	Recent chemistry experiments have strongly indicated that element 114 possesses non eka-lead properties and appears to behave as the first superheavy element to show noble-gas-like properties due to relativistic effects.
	In December 1998, scientists at Dubna (Joint Institute for Nuclear Research) in Russia bombarded a Pu-244 target with Ca-48 ions.
Aqueous solution	An Aqueous solution is a solution in which the solvent is water. It is usually shown in chemical equations by appending (aq) to the relevant formula. The word aqueous means pertaining to, related to, similar to, or dissolved in water.
Solution	In chemistry, a Solution is a homogeneous mixture composed of two or more substances. In such a mixture, a solute is dissolved in another substance, known as a solvent. Gases may dissolve in liquids, for example, carbon dioxide or oxygen in water.
Ball-and-stick model	Ball-and-stick model s and space-filling models are 3D or spatial molecular models which serve to display the structure of chemical products and substances or biomolecules.
	With Ball-and-stick model s, the centers of the atoms are connected by straight lines which represent the covalent bonds. Double and triple bonds are often represented by springs, which form curved connections between the balls.

Chapter 2. Atoms, Molecules, and Ions

Chapter 2. Atoms, Molecules, and Ions

Fullerene	Fullerene is a family of carbon allotropes, molecules composed entirely of carbon, in the form of a hollow sphere, ellipsoid, or tube. Spherical Fullerene s are also called buckyballs, and cylindrical ones are called carbon nanotubes or buckytubes. Fullerene s are similar in structure to graphite, which is composed of stacked sheets of linked hexagonal rings, but may also contain pentagonal (or sometimes heptagonal) rings that would prevent a sheet from being planar.
Helium	Helium is the chemical element with atomic number 2, and is represented by the symbol He. It is a colorless, odorless, tasteless, non-toxic, inert monatomic gas that heads the noble gas group in the periodic table. Its boiling and melting points are the lowest among the elements and it exists only as a gas except in extreme conditions.
Polymer	A Polymer is a large molecule composed of repeating structural units typically connected by covalent chemical bonds. While Polymer in popular usage suggests plastic, the term actually refers to a large class of natural and synthetic materials with a variety of properties. Due to the extraordinary range of properties accessible in Polymer ic materials , they have come to play an essential and ubiquitous role in everyday life - from plastics and elastomers on the one hand to natural bio Polymer s such as DNA and proteins that are essential for life on the other.
Space-filling models	Space-filling models, like ball-and-stick models, belong to the 3D or spatial molecular models, which serve to display the structure of chemical products and substances or biomolecules. This calotte model of one of the many isomers of octane illustrates how the 18 hydrogen atoms line the backbone of 8 carbon atoms The calotte model is a further development of the ball-and-stick model, with which the configuration of molecules can be vividly presented. With it, the atoms of individual elements are represented by multicolored spherical sections.
Sulfur	Sulfur or sulphur is the chemical element that has the atomic number 16. It is denoted with the symbol S. It is an abundant, multivalent non-metal. Sulfur in its native form is a yellow crystalline solid.
Boiling	Boiling, a type of phase transition, is the rapid vaporization of a liquid, which typically occurs when a liquid is heated to its Boiling point, the temperature at which the vapor pressure of the liquid is equal to the pressure exerted on the liquid by the surrounding environmental pressure. Thus, a liquid may also boil when the pressure of the surrounding atmosphere is sufficiently reduced, such as the use of a vacuum pump or at high altitudes. Boiling occurs in three characteristic stages, which are nucleate, transition and film Boiling.
Configuration	The configuration of a molecule is the permanent geometry that results from the spatial arrangement of its bonds. The ability of the same set of atoms to form two or more molecules with different configurations is stereoisomerism. configuration is distinct from chemical conformation, a shape attainable by bond rotations.

Chapter 2. Atoms, Molecules, and Ions

Chapter 2. Atoms, Molecules, and Ions

Crystal	A Crystal or Crystal line solid is a solid material whose constituent atoms, molecules, or ions are arranged in an orderly repeating pattern extending in all three spatial dimensions. The scientific study of Crystal s and Crystal formation is Crystal lography. The process of Crystal formation via mechanisms of Crystal growth is called Crystal lization or solidification.
Crystal structure	In mineralogy and crystallography, a Crystal structure is a unique arrangement of atoms in a crystal. A Crystal structure is composed of a motif, a set of atoms arranged in a particular way, and a lattice. Motifs are located upon the points of a lattice, which is an array of points repeating periodically in three dimensions.
Electron configuration	In atomic physics and quantum chemistry, Electron configuration is the arrangement of electrons of an atom, a molecule, or other physical structure. It concerns the way electrons can be distributed in the orbitals of the given system (atomic or molecular for instance.) Like other elementary particles, the electron is subject to the laws of quantum mechanics, and exhibits both particle-like and wave-like nature.
Enthalpy	In thermodynamics and molecular chemistry, the Enthalpy is a thermodynamic property of a fluid. It can be used to calculate the heat transfer during a quasistatic process taking place in a closed thermodynamic system under constant pressure. Enthalpy H is an arbitrary concept but the Enthalpy change ΔH is more useful because it is equal to the change in the internal energy of the system, plus the work that the system has done on its surroundings.
Melting	Melting is a physical process that results in the phase change of a substance from a solid to a liquid. The internal energy of a solid substance is increased, typically by the application of heat or pressure, resulting in a rise of its temperature to the Melting point, at which the rigid ordering of molecular entities in the solid breaks down to a less-ordered state and the solid liquefies. An object that has melted completely is molten.
Calcium	Calcium is the chemical element with the symbol Ca and atomic number 20. It has an atomic mass of 40.078 amu. Calcium is a soft grey alkaline earth metal, and is the fifth most abundant element by mass in the Earth"s crust.
Ionic compound	In chemistry, an Ionic compound is a chemical compound in which ions are held together in a lattice structure by ionic bonds. Usually, the positively charged portion consists of metal cations and the negatively charged portion is an anion or polyatomic ion. Ions in Ionic compound s are held together by the electrostatic force between oppositely charged bodies.
Peptide bond	A Peptide bond is a chemical bond formed between two molecules when the carboxyl group of one molecule reacts with the amine group of the other molecule, thereby releasing a molecule of water (H_2O.) This is a dehydration synthesis reaction (also known as a condensation reaction), and usually occurs between amino acids. The resulting CO-NH bond is called a Peptide bond, and the resulting molecule is an amide.

Chapter 2. Atoms, Molecules, and Ions

Chapter 2. Atoms, Molecules, and Ions

Base	In chemistry, a Base is most commonly thought of as an aqueous substance that can accept hydrogen ions. A Base is also often referred to as an alkali if OH⁻ ions are involved. This refers to the Brønsted-Lowry theory of acids and bases.
Intermolecular forces	In physics, chemistry, and biology, Intermolecular forces are forces that act between stable molecules or between functional groups of macromolecules. Intermolecular forces include momentary attractions between molecules, diatomic free elements, and individual atoms. They differ from covalent and ionic bonding in that they are not stable, but are caused by momentary polarization of particles.
Iron	Iron is a chemical element with the symbol Fe and atomic number 26. Iron is a group 8 and period 4 element. Iron and Iron alloys (steels) are by far the most common metals and the most common ferromagnetic materials in everyday use.
Formula unit	A Formula unit in chemistry is the empirical formula of an ionic or covalent network solid compound used as an independent entity for stoichiometric calculations. It is the lowest whole number ratio of ions represented in an ionic compound. Examples include ionic NaCl and K_2O and covalent networks such as SiO_2 and C (as diamond or graphite.)
Hydrocarbon	In organic chemistry, a Hydrocarbon is an organic compound consisting entirely of hydrogen and carbon. With relation to chemical terminology, aromatic Hydrocarbon s or arenes, alkanes, alkenes and alkyne-based compounds composed entirely of carbon or hydrogen are referred to as "pure" Hydrocarbon s, whereas other Hydrocarbon s with bonded compounds or impurities of sulfur or nitrogen, are referred to as "impure", and remain somewhat erroneously referred to as Hydrocarbon s. Hydrocarbon s are referred to as consisting of a "backbone" or "skeleton" composed entirely of carbon and hydrogen and other bonded compounds, and have a functional group that generally facilitates combustion.
Organic chemistry	Organic chemistry is a discipline within chemistry which involves the scientific study of the structure, properties, composition, reactions, and preparation (by synthesis or by other means) of chemical compounds that contain carbon. These compounds may contain any number of other elements, including hydrogen, nitrogen, oxygen, the halogens as well as phosphorus, silicon and sulphur. The original definition of "organic" chemistry came from the misconception that organic compounds were always related to life processes.
Uranium	Uranium is a silvery-white metallic chemical element in the actinide series of the periodic table that has the symbol U and atomic number 92. Besides its 92 protons, a Uranium nucleus can have between 141 and 146 neutrons, with 146 and 143 (U-235) in its most common isotopes. The number of electrons in a Uranium atom is 92, 6 of them valence electrons.

Chapter 2. Atoms, Molecules, and Ions

Chapter 2. Atoms, Molecules, and Ions

Chemistry	In the history of science, the etymology of the word Chemistry is a debatable issue. It is agreed that the word "alchemy" is a European one, derived from Arabic, but the origin of the root word, chem, is uncertain. Words similar to it have been found in most ancient languages, with different meanings, but conceivably somehow related to alchemy.
Benzene	Benzene, or benzol, is an organic chemical compound and a known carcinogen with the molecular formula C_6H_6. It is sometimes abbreviated Ph-H. Benzene is a colorless and highly flammable liquid with a sweet smell and a relatively high melting point. Because it is a known carcinogen, its use as an additive in gasoline is now limited, but it is an important industrial solvent and precursor in the production of drugs, plastics, synthetic rubber, and dyes.
Methanation	Methanation is a physical-chemical process to generate Methane from a mixture of various gases out of biomass fermentation or thermo-chemical gasification. The main components are carbon monoxide and hydrogen. The following main process describes the Methanation: $$CO + 3\,H_2 \rightarrow CH_4 + H_2O$$ This process is used for the generation of biogenous natural gas substitute, which can be fed into the gas grid. Methanation is the reverse reaction of steam methane reforming, which converts methane into synthesis gas.
Salt	Salt is a dietary mineral composed primarily of sodium chloride that is essential for animal life, but toxic to most land plants. Salt flavor is one of the basic tastes, an important preservative and a popular food seasoning.
Inorganic compounds	Traditionally, Inorganic compounds are considered to be of a mineral, not biological, origin. Complementarily, most organic compounds are traditionally viewed as being of biological origin. Over the past century, the precise classification of inorganic vs organic compounds has become less important to scientists, primarily because the majority of known compounds are synthetic and not of natural origin.
Monatomic ion	A Monatomic ion is an ion consisting of only one type of element atoms, unlike a polyatomic ion, which consists of more than one type of element in one ion. A type I binary ionic compound contains a metal (cation) that forms only one type of ion. A type II ionic compound contains a metal that forms more than one type of ion, i.e., ions with different charges *note that mercury (I) ions always occur bound together to form Hg_2.
Potassium	Potassium is a chemical element. It has the symbol K, atomic number 19, and atomic mass 39.0983. Potassium was first isolated from potash.
Copper	Copper is a chemical element with the symbol Cu and atomic number 29. It is a ductile metal with very high thermal and electrical conductivity. Pure Copper is rather soft and malleable and a freshly-exposed surface has a pinkish or peachy color.

Chapter 2. Atoms, Molecules, and Ions

Chapter 2. Atoms, Molecules, and Ions

Lead	Lead is a main-group element with symbol Pb and atomic number 82. Lead is a soft, malleable poor metal, also considered to be one of the heavy metals. Lead has a bluish-white color when freshly cut, but tarnishes to a dull grayish color when exposed to air.
Zinc	Zinc is a metallic chemical element with the symbol Zn and atomic number 30. It is a first-row transition metal in group 12 of the periodic table. Zinc is chemically similar to magnesium because its ion is of similar size and its only common oxidation state is +2.
Oxidation number	The Oxidation number of a central atom in a coordination compound is the charge that it would have if all the ligands were removed along with the electron pairs that were shared with the central atom. It is used in the nomenclature of inorganic compounds. It is represented by a Roman numeral; the plus sign is omitted for positive Oxidation number s.
Nitrogen	Nitrogen is a chemical element that has the symbol N and atomic number 7 and atomic mass 14.00674 u. Elemental Nitrogen is a colorless, odorless, tasteless and mostly inert diatomic gas at standard conditions, constituting 78% by volume of Earth"s atmosphere. Many industrially important compounds, such as ammonia, nitric acid, organic nitrates , and cyanides, contain Nitrogen.
Polyaniline	Polyaniline is a conducting polymer of the semi-flexible rod polymer family. Although it was discovered over 150 years ago, only recently has Polyaniline captured the attention of the scientific community due to the discovery of its high electrical conductivity. Amongst the family of conducting polymers, Polyaniline is unique due to its ease of synthesis, environmental stability, and simple doping/dedoping chemistry.
Polyatomic ion	A Polyatomic ion is a charged species composed of two or more atoms covalently bonded or of a metal complex that can be considered as acting as a single unit in the context of acid and base chemistry or in the formation of salts. The prefix poly- means many in Greek, but even ions of two atoms are commonly referred to as polyatomic. In older literature, a Polyatomic ion is also referred to as a radical, and less commonly, as a radical group.
Magnesium	Magnesium is a chemical element with the symbol Mg, atomic number 12, atomic weight 24.3050 and common oxidation number +2. Magnesium, an alkaline earth metal, is the ninth most abundant element in the universe by mass. The commonness of Magnesium is related to the fact that it is easily built up in supernova stars from a sequential addition of three helium nuclei to carbon .
Mercury	There are seven isotopes of mercury with Hg-202 being the most abundant (29.86%.) The longest-lived radioisotopes are ^{194}Hg with a half-life of 444 years, and ^{203}Hg with a half-life of 46.612 days. Most of the remaining radioisotopes have half-lives that are less than a day.

Chapter 2. Atoms, Molecules, and Ions

Chapter 2. Atoms, Molecules, and Ions

Titanium	Titanium is a chemical element with the symbol Ti and atomic number 22. Sometimes called the "space age metal", it has a low density and is a strong, lustrous, corrosion-resistant transition metal with a silver color. Titanium can be alloyed with iron, aluminium, vanadium, molybdenum, among other elements, to produce strong lightweight alloys for aerospace (jet engines, missiles, and spacecraft), military, industrial process (chemicals and petro-chemicals, desalination plants, pulp, and paper), automotive, agri-food, medical prostheses, orthopedic implants, dental and endodontic instruments and files, dental implants, sporting goods, jewelry, mobile phones, and other applications.
Covalent radius	The Covalent radius, r_{cov}, is a measure of the size of atom that forms part of a covalent bond. It is measured either in picometres (pm) or ångströms (Å), with 1 Å = 100 pm. In principle, the sum of the two covalent radii should equal the covalent bond length between two atoms.
Dipole	In physics, there are two kinds of Dipole s : - An electric Dipole is a separation of positive and negative charge. The simplest example of this is a pair of electric charges of equal magnitude but opposite sign, separated by some, usually small, distance. A permanent electric Dipole is called an electret. - A magnetic Dipole is a closed circulation of electric current. A simple example of this is a single loop of wire with some constant current flowing through it. Dipole s can be characterized by their Dipole moment, a vector quantity. For the simple electric Dipole given above, the electric Dipole moment would point from the negative charge towards the positive charge, and have a magnitude equal to the strength of each charge times the separation between the charges. For the current loop, the magnetic Dipole moment would point through the loop (according to the right hand grip rule), with a magnitude equal to the current in the loop times the area of the loop. In addition to current loops, the electron, among other fundamental particles, is said to have a magnetic Dipole moment.
Hydrogen sulfide	Hydrogen sulfide is the chemical compound with the formula H_2S. This colorless, toxic and flammable gas is partially responsible for the foul odor of rotten eggs and flatulence. It often results from the bacterial break down of sulfites in nonorganic matter in the absence of oxygen, such as in swamps and sewers (anaerobic digestion.) It also occurs in volcanic gases, natural gas and some well waters.
Nitric acid	Nitric acid, also known as aqua fortis and spirit of nitre, is a highly corrosive and toxic strong acid that can cause severe burns. Colorless when pure, older samples tend to acquire a yellow cast due to the accumulation of oxides of nitrogen. If the solution contains more than 86% Nitric acid, it is referred to as fuming Nitric acid.

Chapter 2. Atoms, Molecules, and Ions

Chapter 2. Atoms, Molecules, and Ions

Anhydrous	As a general term, a substance is said to be Anhydrous if it contains no water. The way of achieving the Anhydrous form differs from one substance to another. In many cases, the presence of water can prevent a reaction from happening, or form undesirable products.
Hydrobromic acid	Hydrobromic acid is a strong acid formed by dissolving the diatomic molecule hydrogen bromide in water. "Constant boiling" Hydrobromic acid is an aqueous solution that distills at 124.3 °C and contains 47.6% HBr by weight. Hydrobromic acid has a pK_a of −9, making it a stronger acid than hydrochloric acid, but not as strong as hydroiodic acid.
Hydrochloric acid	Hydrochloric acid is the solution of hydrogen chloride (HCl) in water. It is a highly corrosive, strong mineral acid and has major industrial uses. It is found naturally in gastric acid.
Hydrofluoric acid	Hydrofluoric acid is a solution of hydrogen fluoride in water. While it is extremely corrosive and dangerous to handle, it is technically a weak acid. Hydrogen fluoride, often in the aqueous form as Hydrofluoric acid, is a valued source of fluorine, being the precursor to numerous pharmaceuticals (e.g., Prozac), diverse polymers (e.g., Teflon), and most other synthetic materials that contain fluorine.
Hydrogen bromide	Hydrogen bromide is the diatomic molecule Hydrogen bromide r. Hydrogen bromide r is a gas that liquifies upon cooling. Hydrobromic acid forms upon dissolving Hydrogen bromide r in water.
Electrolysis	In chemistry and manufacturing, Electrolysis is a method of using an electric current to drive an otherwise non-spontaneous chemical reaction. Electrolysis is commercially highly important as a stage in the separation of elements from naturally-occurring sources such as ores using an electrolytic cell. - 1800 - William Nicholson and Johann Ritter decomposed water into hydrogen and oxygen. - 1807 - Potassium, Sodium, Barium, Calcium and Magnesium were discovered by Sir Humphry Davy using Electrolysis. - 1886 - Fluorine was discovered by Henri Moissan using Electrolysis. - 1886 - Hall-Héroult process developed for making aluminium - 1890 - Castner-Kellner process developed for making sodium hydroxide Electrolysis is the passage of an electric current through an ionic substance that is either molten or dissolved in a suitable solvent, resulting in chemical reactions at the electrodes and separation of materials.

Chapter 2. Atoms, Molecules, and Ions

Chapter 2. Atoms, Molecules, and Ions

	The main components required to achieve Electrolysis are: • A liquid containing mobile ions - an electrolyte • An external source of direct electric current • Two solid rods or plates known as electrodes The components perform the following roles in the Electrolysis process: • The mobile ions are the carriers of electrical current in the liquid (electrolyte.) If the ions are not mobile, as in a solid salt then Electrolysis cannot occur.
Chemical equation	A Chemical equation may be described as a chemical reaction or a means of writing out and describing such a phenomenon. The coefficients next to the symbols and formulae of entities are the absolute values of the stoichiometric numbers. The first Chemical equation was diagrammed by Jean Beguin in 1615.
Catalyst	Catalyst is a science centre and museum devoted to the chemical industry. Its full title is Catalyst Science Discovery Centre. It is located in Widnes, Cheshire, in the north west of England, and situated on the north bank of the River Mersey (grid reference SJ512841.)
Platinum	Platinum is a chemical element with the chemical symbol Pt and an atomic number of 78." It is in Group 10 of the periodic table of elements. A dense, malleable, ductile, precious, gray-white transition metal, Platinum is resistant to corrosion and occurs in some nickel and copper ores along with some native deposits. Platinum is used in jewelry, laboratory equipment, electrical contacts and electrodes, Platinum resistance thermometers, dentistry equipment, and catalytic converters.
Frequency spectrum	A source of light can have many colors mixed together and in different amounts (intensities.) A rainbow sends the different frequencies in different directions, making them individually visible at different angles. A graph of the intensity plotted against the frequency (showing the amount of each color) is the Frequency spectrum of the light.
Product	A Product is a substance that forms as a result of a biological- or chemical reaction. While the end Product of some chemical reactions may be the result of a relatively rapid reaction, nanoseconds to seconds, chemical equilibria in complex systems may require years or even centuries to be established. For example, equilibria in groundwater systems with multiple components are achieved on timescales of millennia, if ever.
Sodium nitrate	Sodium nitrate is the chemical compound with the formula $NaNO_3$. This salt, also known as "Chile saltpeter" or "Peru saltpeter" (to distinguish it from ordinary saltpeter, potassium nitrate), is a white solid which is very soluble in water. The mineral form is also known as nitratine or soda niter.

Chapter 2. Atoms, Molecules, and Ions

Chapter 2. Atoms, Molecules, and Ions

Yield | Nuclear fission splits a heavy nucleus such as uranium or plutonium into two lighter nuclei, which are called fission products. Yield refers to the fraction of a fission product produced per fission. Yield can be broken down by:

1. Individual isotope
2. Chemical element spanning several isotopes of different mass number but same atomic number.
3. Nuclei of a given mass number regardless of atomic number. Known as "chain Yield" because it represents a decay chain of beta decay.

Isotope and element yields will change as the fission products undergo beta decay, while chain yields do not change after completion of neutron emission by a few neutron-rich initial fission products (delayed neutrons), with halflife measured in seconds.

A few isotopes can be produced directly by fission, but not by beta decay because the would-be precursor with atomic number one greater is stable and does not decay.

Chapter 2. Atoms, Molecules, and Ions

Chapter 3. Calculations with Chemical Formulas and Equations

Atomic mass	The Atomic mass is the mass of an atom, most often expressed in unified Atomic mass units. The Atomic mass may be considered to be the total mass of protons, neutrons and electrons in a single atom (when the atom is motionless.) The Atomic mass is sometimes incorrectly used as a synonym of relative Atomic mass, average Atomic mass and atomic weight; however, these differ subtly from the Atomic mass.
Molecular mass	The Molecular mass of a substance, frequently referred by the older term molecular weight and abbreviated as MW, is the mass of one molecule of that substance, relative to the unified atomic mass unit u (equal to 1/12 the mass of one isotope of carbon-12.) This is distinct from the relative Molecular mass of a molecule, which is the ratio of the mass of that molecule to 1/12 of the mass of carbon 12 and is a dimensionless number. Relative Molecular mass is abbreviated to M_r.
Oxygen	Oxygen and -γενÎ®ς (-genÄ"s) (producer, literally begetter) is the element with atomic number 8 and represented by the symbol O. It is a member of the chalcogen group on the periodic table, and is a highly reactive nonmetallic period 2 element that readily forms compounds (notably oxides) with almost all other elements. At standard temperature and pressure two atoms of the element bind to form di Oxygen, a colorless, odorless, tasteless diatomic gas with the formula O_2. Oxygen is the third most abundant element in the universe by mass after hydrogen and helium and the most abundant element by mass in the Earth"s crust.
Acid	An Acid is traditionally considered any chemical compound that, when dissolved in water, gives a solution with a hydrogen ion activity greater than in pure water, i.e. a pH less than 7.0. That approximates the modern definition of Johannes Nicolaus Brønsted and Martin Lowry, who independently defined an Acid as a compound which donates a hydrogen ion (H^+) to another compound (called a base.) Common examples include acetic Acid and sulfuric Acid (used in car batteries.)
Ethanol	Ethanol pure alcohol, grain alcohol is a volatile, flammable, colorless liquid. It is a psychoactive drug, best known as the type of alcohol found in alcoholic beverages and in modern thermometers. Ethanol is one of the oldest recreational drugs.
Freezing	In physical science, Freezing or solidification is the process in which a liquid turns into a solid when cold enough. The Freezing point is the temperature at which this happens. Melting, the process of turning a solid to a liquid, is almost the exact opposite of Freezing.
Melting	Melting is a physical process that results in the phase change of a substance from a solid to a liquid. The internal energy of a solid substance is increased, typically by the application of heat or pressure, resulting in a rise of its temperature to the Melting point, at which the rigid ordering of molecular entities in the solid breaks down to a less-ordered state and the solid liquefies. An object that has melted completely is molten.

Chapter 3. Calculations with Chemical Formulas and Equations

Chapter 3. Calculations with Chemical Formulas and Equations

Iron	Iron is a chemical element with the symbol Fe and atomic number 26. Iron is a group 8 and period 4 element. Iron and Iron alloys (steels) are by far the most common metals and the most common ferromagnetic materials in everyday use.
Molecular model	A Molecular model in this article, is a physical model that represents molecules and their processes. The creation of mathematical models of molecular properties and behaviour is Molecular model ling, and their graphical depiction is molecular graphics, but these topics are closely linked and each uses techniques from the others. In this article, Molecular model will primarily refer to systems containing more than one atom and where nuclear structure is neglected.
Sodium	Sodium is a metallic element with a symbol Na and atomic number 11. It is a soft, silvery-white, highly reactive metal and is a member of the alkali metals within "group 1" (formerly known as "group IA".) It has only one stable isotope, ^{23}Na.
Water	Water is the most abundant molecule on Earth"s surface, constituting about 75% of the Earth"s surface. In nature it exists in liquid, solid, and gaseous states. It is in dynamic equilibrium between the liquid and gas states at standard temperature and pressure.
Chlorination	Chlorination is the process of adding the element chlorine to water as a method of water purification to make it fit for human consumption as drinking water. Water which has been treated with chlorine is effective in preventing the spread of disease. The Chlorination of public drinking supplies was originally met with resistance, as people were concerned about the health effects of the practice.
Carbon	Carbon is the chemical element with symbol C and atomic number 6. As a member of group 14 on the periodic table, it is nonmetallic and tetravalent--making four electrons available to form covalent chemical bonds. There are three naturally occurring isotopes, with ^{12}C and ^{13}C being stable, while ^{14}C is radioactive, decaying with a half-life of about 5730 years.
Mole	The mole is a unit of amount of substance: it is an SI base unit, and one of the few units used to measure this physical quantity. The name "mole" was coined in German by Wilhelm Ostwald in 1893, although the related concept of equivalent mass had been in use at least a century earlier. The name is assumed to be derived from the word Molekül (molecule.)
Moles	Moles are the majority of the members of the mammal family Talpidae in the order Soricomorpha. Although most moles burrow, some species are aquatic or semi-aquatic. moles have cylindrical bodies covered in fur, with small or covered eyes; the ears are generally not visible.

Chapter 3. Calculations with Chemical Formulas and Equations

Chapter 3. Calculations with Chemical Formulas and Equations

Chemical formula	A Chemical formula is a way of expressing information about the atoms that constitute a particular chemical compound, and how the relationship between those atoms changes in chemical reactions. For molecular compounds it is also known as the molecular formula, and identifies each constituent element by its chemical symbol and indicates the number of atoms of each element found in each discrete molecule of that compound. If a molecule contains more than one atom of a particular element, this quantity is indicated using a subscript after the chemical symbol (although 19th-century books often used superscripts.)
Hydrogen	Hydrogen is the chemical element with atomic number 1. It is represented by the symbol H. At standard temperature and pressure, Hydrogen is a colorless, odorless, nonmetallic, tasteless, highly flammable diatomic gas with the molecular formula H_2. With an atomic weight of 1.007 94 u, Hydrogen is the lightest element.
Molar mass	Molar mass symbol M, is the mass of one mole of a substance (chemical element or chemical compound.) It is a physical property which is characteristic of each pure substance. The base SI unit for mass is the kilogram but, for both practical and historical reasons, Molar mass es are almost always quoted in grams per mole (g/mol or g mol^{-1}), especially in chemistry.
Atom	The Atom is a basic unit of matter consisting of a dense, central nucleus surrounded by a cloud of negatively charged electrons. The atomic nucleus contains a mix of positively charged protons and electrically neutral neutrons (except in the case of hydrogen-1, which is the only stable nuclide with no neutron.) The electrons of an Atom are bound to the nucleus by the electromagnetic force.
Molecule	A Molecule is defined as a sufficiently stable, electrically neutral group of at least two atoms in a definite arrangement held together by very strong (covalent) chemical bonds. Molecule s are distinguished from polyatomic ions in this strict sense. In organic chemistry and biochemistry, the term Molecule is used less strictly and also is applied to charged organic Molecule s and bio Molecule s.
Zinc	Zinc is a metallic chemical element with the symbol Zn and atomic number 30. It is a first-row transition metal in group 12 of the periodic table. Zinc is chemically similar to magnesium because its ion is of similar size and its only common oxidation state is +2.
Iodine	Iodine , is a chemical element that has the symbol I and atomic number 53. Naturally-occurring Iodine is a single isotope with 74 neutrons. Chemically, Iodine is the second least reactive of the halogens, and the second most electropositive halogen; trailing behind astatine in both of these categories.
Lead	Lead is a main-group element with symbol Pb and atomic number 82. Lead is a soft, malleable poor metal, also considered to be one of the heavy metals. Lead has a bluish-white color when freshly cut, but tarnishes to a dull grayish color when exposed to air.

Chapter 3. Calculations with Chemical Formulas and Equations

Chapter 3. Calculations with Chemical Formulas and Equations

Combustion	Combustion or burning is a complex sequence of exothermic chemical reactions between a fuel (usually a hydrocarbon) and an oxidant accompanied by the production of heat or both heat and light in the form of either a glow or flames, appearance of light flickering. Direct Combustion by atmospheric oxygen is a reaction mediated by radical intermediates. The conditions for radical production are naturally produced by thermal runaway, where the heat generated by Combustion is necessary to maintain the high temperature necessary for radical production.
Elemental analysis	Elemental analysis is a process where a sample of some material (e.g., soil, waste or drinking water, bodily fluids, minerals, chemical compounds) is analyzed for its elemental and sometimes isotopic composition. Elemental analysis can be qualitative (determining what elements are present), and it can be quantitative (determining how much of each are present.) Elemental analysis falls within the ambit of analytical chemistry, the set of instruments involved in deciphering the chemical nature of our world.
Boiling	Boiling, a type of phase transition, is the rapid vaporization of a liquid, which typically occurs when a liquid is heated to its Boiling point, the temperature at which the vapor pressure of the liquid is equal to the pressure exerted on the liquid by the surrounding environmental pressure. Thus, a liquid may also boil when the pressure of the surrounding atmosphere is sufficiently reduced, such as the use of a vacuum pump or at high altitudes. Boiling occurs in three characteristic stages, which are nucleate, transition and film Boiling.
Benzene	Benzene, or benzol, is an organic chemical compound and a known carcinogen with the molecular formula C_6H_6. It is sometimes abbreviated Ph-H. Benzene is a colorless and highly flammable liquid with a sweet smell and a relatively high melting point. Because it is a known carcinogen, its use as an additive in gasoline is now limited, but it is an important industrial solvent and precursor in the production of drugs, plastics, synthetic rubber, and dyes.
Empirical formula	In chemistry, the Empirical formula of a chemical compound is a simple expression of the relative numbers of each type of atom in it, or the simplest whole number ratio of atoms of each element present in a compound. An Empirical formula makes no reference to isomerism, structure, or absolute number of atoms. The Empirical formula is used as standard for most ionic compounds, such as $CaCl_2$, and for macromolecules, such as SiO_2.
Mass spectrum	A Mass spectrum is an intensity vs. m/z (mass-to-charge ratio) plot representing a chemical analysis. Hence, the Mass spectrum of a sample is a pattern representing the distribution of components (atoms or molecules) by mass (more correctly: mass-to-charge ratio) in a sample. It is usually acquired using an instrument called a mass spectrometer.
Chromium	Chromium is a chemical element which has the symbol Cr and atomic number 24. It is a steely-gray, lustrous, hard metal that takes a high polish and has a high melting point. It is also odourless, tasteless, and malleable.

Chapter 3. Calculations with Chemical Formulas and Equations

Chapter 3. Calculations with Chemical Formulas and Equations

Stoichiometry	Stoichiometry is the calculation of quantitative (measurable) relationships of the reactants and products in a balanced chemical reaction (chemicals.) It can be used to calculate quantities such as the amount of products that can be produced with the given reactants and percent yield. "Stoichiometry" is derived from the Greek words στοιχεά¿–ον and μÎτρον (metron, meaning measure.)
Acid-base reaction	An Acid-base reaction is a chemical reaction that occurs between an acid and a base. Several concepts that provide alternative definitions for the reaction mechanisms involved and their application in solving related problems exist. Despite several differences in definitions, their importance becomes apparent as different methods of analysis when applied to Acid-base reaction s for gaseous or liquid species, or when acid or base character may be somewhat less apparent.
Ammonia	Ammonia is a compound of nitrogen and hydrogen with the formula NH_3. It is normally encountered as a gas with a characteristic pungent odor. Ammonia contributes significantly to the nutritional needs of terrestrial organisms by serving as a precursor to foodstuffs and fertilizers.
Chemical equation	A Chemical equation may be described as a chemical reaction or a means of writing out and describing such a phenomenon. The coefficients next to the symbols and formulae of entities are the absolute values of the stoichiometric numbers. The first Chemical equation was diagrammed by Jean Beguin in 1615.
Haber process	The Haber process is the nitrogen fixation reaction of nitrogen gas and hydrogen gas, over an enriched iron catalyst, to produce ammonia. The Haber process is important because ammonia is difficult to produce on an industrial scale, and the fertilizer generated from the ammonia is responsible for sustaining one-third of the Earth"s population. Despite the fact that 78.1% of the air we breathe is nitrogen, the gas is relatively unreactive because nitrogen molecules are held together by strong triple bonds.
Nitrogen	Nitrogen is a chemical element that has the symbol N and atomic number 7 and atomic mass 14.00674 u. Elemental Nitrogen is a colorless, odorless, tasteless and mostly inert diatomic gas at standard conditions, constituting 78% by volume of Earth"s atmosphere. Many industrially important compounds, such as ammonia, nitric acid, organic nitrates , and cyanides, contain Nitrogen.
Aqueous solution	An Aqueous solution is a solution in which the solvent is water. It is usually shown in chemical equations by appending (aq) to the relevant formula. The word aqueous means pertaining to, related to, similar to, or dissolved in water.
Solution	In chemistry, a Solution is a homogeneous mixture composed of two or more substances. In such a mixture, a solute is dissolved in another substance, known as a solvent. Gases may dissolve in liquids, for example, carbon dioxide or oxygen in water.

Chapter 3. Calculations with Chemical Formulas and Equations

Chapter 3. Calculations with Chemical Formulas and Equations

Amount of substance	In physical sciences, the Amount of substance, n, of a sample can be defined informally as the number of some specified elementary entities (usually either atoms or ions but where this number is expressed in terms of some standard batch size. This is analogous to expressing the number of years in terms of centuries see below for a formal definition. What makes amount-of-substance a useful quantity--even though it may seem that it is a redundant one, given that it seems to contain the same information as does the total number of entities--is its particular convenience in real-world experimental settings, a convenience that largely stems from the specific choice that was made for the definition of the standard batch size.
Frequency spectrum	A source of light can have many colors mixed together and in different amounts (intensities.) A rainbow sends the different frequencies in different directions, making them individually visible at different angles. A graph of the intensity plotted against the frequency (showing the amount of each color) is the Frequency spectrum of the light.
Product	A Product is a substance that forms as a result of a biological- or chemical reaction. While the end Product of some chemical reactions may be the result of a relatively rapid reaction, nanoseconds to seconds, chemical equilibria in complex systems may require years or even centuries to be established. For example, equilibria in groundwater systems with multiple components are achieved on timescales of millennia, if ever.
Yield	Nuclear fission splits a heavy nucleus such as uranium or plutonium into two lighter nuclei, which are called fission products. Yield refers to the fraction of a fission product produced per fission. Yield can be broken down by: 1. Individual isotope 2. Chemical element spanning several isotopes of different mass number but same atomic number. 3. Nuclei of a given mass number regardless of atomic number. Known as "chain Yield" because it represents a decay chain of beta decay. Isotope and element yields will change as the fission products undergo beta decay, while chain yields do not change after completion of neutron emission by a few neutron-rich initial fission products (delayed neutrons), with halflife measured in seconds. A few isotopes can be produced directly by fission, but not by beta decay because the would-be precursor with atomic number one greater is stable and does not decay.
Hydrochloric acid	Hydrochloric acid is the solution of hydrogen chloride (HCl) in water. It is a highly corrosive, strong mineral acid and has major industrial uses. It is found naturally in gastric acid.
Mercury	There are seven isotopes of mercury with Hg-202 being the most abundant (29.86%.) The longest-lived radioisotopes are ^{194}Hg with a half-life of 444 years, and ^{203}Hg with a half-life of 46.612 days. Most of the remaining radioisotopes have half-lives that are less than a day.

Chapter 3. Calculations with Chemical Formulas and Equations

Chapter 3. Calculations with Chemical Formulas and Equations

Mercury oxide	Mercury oxide has a formula of HgO and a formula weight of 216.6. It has a red or orange color. Mercury oxide is a solid at room temperature and pressure.
Joseph Priestley	Joseph Priestley - 6 February 1804) was an 18th-century British theologian, Dissenting clergyman, natural philosopher, educator, and political theorist who published over 150 works. He is usually credited with the discovery of oxygen, having isolated it in its gaseous state, although Carl Wilhelm Scheele and Antoine Lavoisier also have a claim to the discovery. During his lifetime, Priestley"s considerable scientific reputation rested on his invention of soda water, his writings on electricity, and his discovery of several "airs" , the most famous being what Priestley dubbed "dephlogisticated air" (oxygen.)
Crystal	A Crystal or Crystal line solid is a solid material whose constituent atoms, molecules, or ions are arranged in an orderly repeating pattern extending in all three spatial dimensions. The scientific study of Crystal s and Crystal formation is Crystal lography. The process of Crystal formation via mechanisms of Crystal growth is called Crystal lization or solidification.
Crystal structure	In mineralogy and crystallography, a Crystal structure is a unique arrangement of atoms in a crystal. A Crystal structure is composed of a motif, a set of atoms arranged in a particular way, and a lattice. Motifs are located upon the points of a lattice, which is an array of points repeating periodically in three dimensions.
Sulfur	Sulfur or sulphur is the chemical element that has the atomic number 16. It is denoted with the symbol S. It is an abundant, multivalent non-metal. Sulfur in its native form is a yellow crystalline solid.

Chapter 3. Calculations with Chemical Formulas and Equations

Chapter 4. Chemical Reactions

Lead	Lead is a main-group element with symbol Pb and atomic number 82. Lead is a soft, malleable poor metal, also considered to be one of the heavy metals. Lead has a bluish-white color when freshly cut, but tarnishes to a dull grayish color when exposed to air.
Potassium	Potassium is a chemical element. It has the symbol K, atomic number 19, and atomic mass 39.0983. Potassium was first isolated from potash.
Precipitation	Precipitation is the formation of a solid in a solution during a chemical reaction. When the reaction occurs, the solid formed is called the precipitate, and the liquid remaining above the solid is called the supernate. Natural methods of Precipitation include settling or sedimentation, where a solid forms over a period of time due to ambient forces like gravity or centrifugation.
Solution	In chemistry, a Solution is a homogeneous mixture composed of two or more substances. In such a mixture, a solute is dissolved in another substance, known as a solvent. Gases may dissolve in liquids, for example, carbon dioxide or oxygen in water.
Acid-base reaction	An Acid-base reaction is a chemical reaction that occurs between an acid and a base. Several concepts that provide alternative definitions for the reaction mechanisms involved and their application in solving related problems exist. Despite several differences in definitions, their importance becomes apparent as different methods of analysis when applied to Acid-base reaction s for gaseous or liquid species, or when acid or base character may be somewhat less apparent.
Concentration	In chemistry, Concentration is the measure of how much of a given substance there is mixed with another substance. This can apply to any sort of chemical mixture, but most frequently the concept is limited to homogeneous solutions, where it refers to the amount of solute in the solvent. To concentrate a solution, one must add more solute, or reduce the amount of solvent (for instance, by selective evaporation.)
Electrolysis	In chemistry and manufacturing, Electrolysis is a method of using an electric current to drive an otherwise non-spontaneous chemical reaction. Electrolysis is commercially highly important as a stage in the separation of elements from naturally-occurring sources such as ores using an electrolytic cell. - 1800 - William Nicholson and Johann Ritter decomposed water into hydrogen and oxygen. - 1807 - Potassium, Sodium, Barium, Calcium and Magnesium were discovered by Sir Humphry Davy using Electrolysis. - 1886 - Fluorine was discovered by Henri Moissan using Electrolysis. - 1886 - Hall-Héroult process developed for making aluminium - 1890 - Castner-Kellner process developed for making sodium hydroxide

Chapter 4. Chemical Reactions

Chapter 4. Chemical Reactions

Electrolysis is the passage of an electric current through an ionic substance that is either molten or dissolved in a suitable solvent, resulting in chemical reactions at the electrodes and separation of materials.

The main components required to achieve Electrolysis are:

- A liquid containing mobile ions - an electrolyte
- An external source of direct electric current
- Two solid rods or plates known as electrodes

The components perform the following roles in the Electrolysis process:

- The mobile ions are the carriers of electrical current in the liquid (electrolyte.) If the ions are not mobile, as in a solid salt then Electrolysis cannot occur.

Graphite	Graphite " href="/wiki/Cleavage_(crystal)">loose interlamellar coupling between sheets in the structure, in fact in a vacuum environment (such as in technologies for use in space), Graphite was found to be a very poor lubricant. This fact led to the discovery that Graphite"s lubricity is due to adsorbed air and water between the layers, unlike other layered dry lubricants such as molybdenum disulfide. Recent studies suggest that an effect called superlubricity can also account for this effect.
Hydrochloric acid	Hydrochloric acid is the solution of hydrogen chloride (HCl) in water. It is a highly corrosive, strong mineral acid and has major industrial uses. It is found naturally in gastric acid.
Hydrogen	Hydrogen is the chemical element with atomic number 1. It is represented by the symbol H. At standard temperature and pressure, Hydrogen is a colorless, odorless, nonmetallic, tasteless, highly flammable diatomic gas with the molecular formula H_2. With an atomic weight of 1.007 94 u, Hydrogen is the lightest element.
Sodium	Sodium is a metallic element with a symbol Na and atomic number 11. It is a soft, silvery-white, highly reactive metal and is a member of the alkali metals within "group 1" (formerly known as "group IA".) It has only one stable isotope, ^{23}Na.
Sucrose	Sucrose is a disaccharide of glucose and fructose with an α 1,2 glycosidic linkage. The molecular formula of Sucrose is $C_{12}H_{22}O_{11}$. Its systematic name is β-D-fructofuranosyl--α-D-glucopyranoside
Water	Water is the most abundant molecule on Earth"s surface, constituting about 75% of the Earth"s surface. In nature it exists in liquid, solid, and gaseous states. It is in dynamic equilibrium between the liquid and gas states at standard temperature and pressure.

Chapter 4. Chemical Reactions

Chapter 4. Chemical Reactions

Acid	An Acid is traditionally considered any chemical compound that, when dissolved in water, gives a solution with a hydrogen ion activity greater than in pure water, i.e. a pH less than 7.0. That approximates the modern definition of Johannes Nicolaus Brønsted and Martin Lowry, who independently defined an Acid as a compound which donates a hydrogen ion (H^+) to another compound (called a base.) Common examples include acetic Acid and sulfuric Acid (used in car batteries.)
Aqueous Solution	An Aqueous solution is a solution in which the solvent is water. It is usually shown in chemical equations by appending (aq) to the relevant formula. The word aqueous means pertaining to, related to, similar to, or dissolved in water.
Crystal	A Crystal or Crystal line solid is a solid material whose constituent atoms, molecules, or ions are arranged in an orderly repeating pattern extending in all three spatial dimensions. The scientific study of Crystal s and Crystal formation is Crystal lography. The process of Crystal formation via mechanisms of Crystal growth is called Crystal lization or solidification.
Crystal structure	In mineralogy and crystallography, a Crystal structure is a unique arrangement of atoms in a crystal. A Crystal structure is composed of a motif, a set of atoms arranged in a particular way, and a lattice. Motifs are located upon the points of a lattice, which is an array of points repeating periodically in three dimensions.
Impurities	Impurities are substances inside a confined amount of liquid, gas which differ from the chemical composition of the material or compound. Impurities are either naturally occurring or added during synthesis of a chemical or commercial product. During production, Impurities may be purposely, accidentally, inevitably, or incidentally added into the substance.
Ammonia	Ammonia is a compound of nitrogen and hydrogen with the formula NH_3. It is normally encountered as a gas with a characteristic pungent odor. Ammonia contributes significantly to the nutritional needs of terrestrial organisms by serving as a precursor to foodstuffs and fertilizers.
Base	In chemistry, a Base is most commonly thought of as an aqueous substance that can accept hydrogen ions. A Base is also often referred to as an alkali if OH^- ions are involved. This refers to the Brønsted-Lowry theory of acids and bases.
Benzene	Benzene, or benzol, is an organic chemical compound and a known carcinogen with the molecular formula C_6H_6. It is sometimes abbreviated Ph-H. Benzene is a colorless and highly flammable liquid with a sweet smell and a relatively high melting point. Because it is a known carcinogen, its use as an additive in gasoline is now limited, but it is an important industrial solvent and precursor in the production of drugs, plastics, synthetic rubber, and dyes.

Chapter 4. Chemical Reactions

Chapter 4. Chemical Reactions

Calcium	Calcium is the chemical element with the symbol Ca and atomic number 20. It has an atomic mass of 40.078 amu. Calcium is a soft grey alkaline earth metal, and is the fifth most abundant element by mass in the Earth"s crust.
Ionic compound	In chemistry, an Ionic compound is a chemical compound in which ions are held together in a lattice structure by ionic bonds. Usually, the positively charged portion consists of metal cations and the negatively charged portion is an anion or polyatomic ion. Ions in Ionic compound s are held together by the electrostatic force between oppositely charged bodies.
Acid rain	Acid rain is rain or any other form of precipitation that is unusually acidic. It has harmful effects on plants, aquatic animals, and infrastructure. Acid rain is mostly caused by human emissions of sulfur and nitrogen compounds which react in the atmosphere to produce acids.
Molecular model	A Molecular model in this article, is a physical model that represents molecules and their processes. The creation of mathematical models of molecular properties and behaviour is Molecular model ling, and their graphical depiction is molecular graphics, but these topics are closely linked and each uses techniques from the others. In this article, Molecular model will primarily refer to systems containing more than one atom and where nuclear structure is neglected.
Mercury	There are seven isotopes of mercury with Hg-202 being the most abundant (29.86%.) The longest-lived radioisotopes are ^{194}Hg with a half-life of 444 years, and ^{203}Hg with a half-life of 46.612 days. Most of the remaining radioisotopes have half-lives that are less than a day.
Chemical equation	A Chemical equation may be described as a chemical reaction or a means of writing out and describing such a phenomenon. The coefficients next to the symbols and formulae of entities are the absolute values of the stoichiometric numbers. The first Chemical equation was diagrammed by Jean Beguin in 1615.
Calcium hydroxide	Calcium hydroxide, traditionally called slaked lime, hydrated lime, slack lime or pickling lime, is a chemical compound with the chemical formula $Ca(OH)_2$. It is a colourless crystal or white powder, and is obtained when calcium oxide (called lime or quicklime) is mixed, or "slaked" with water. It can also be precipitated by mixing an aqueous solution of calcium chloride and an aqueous solution of sodium hydroxide.
Hydroxide	In chemistry, Hydroxide is the name for the diatomic anion OH^-, consisting of oxygen and hydrogen atoms, usually derived from the dissociation of a base. It is one of the simplest diatomic ions known. Inorganic compounds that contain the hydroxyl group are referred to as Hydroxide s.
Ion	An Ion is an atom or molecule where the total number of electrons is not equal to the total number of protons, giving it a net positive or negative electrical charge.

Chapter 4. Chemical Reactions

Chapter 4. Chemical Reactions

Since protons are positively charged and electrons are negatively charged, if there are more electrons than protons, the atom or molecule will be negatively charged. This is called an an Ion, from the Greek á¼€vÎ¬, meaning "up".

Sodium hydroxide

Sodium hydroxide is a caustic metallic base. Sodium hydroxide forms a strong alkaline solution when dissolved in a solvent such as water. However, only the hydroxide ion is basic.

Perchloric acid

Perchloric acid, $HClO_4$, is an oxoacid of chlorine and is a colorless liquid soluble in water. It is a strong acid comparable in strength to sulfuric and nitric acids. It is useful for preparing perchlorate salts, but it is also dangerously corrosive and readily forms explosive mixtures.

Potassium nitrate

Potassium nitrate is a chemical compound with the chemical formula KNO_3. A naturally occurring mineral source of nitrogen, KNO_3 constitutes a critical oxidizing component of black powder/gunpowder. In the past it was also used for several kinds of burning fuses, including slow matches.

Seawater

Seawater is water from a sea or ocean. On average, Seawater in the world"s oceans has a salinity of about 3.5%. This means that every 1 kg of Seawater has approximately 35 grams of dissolved salts (mostly, but not entirely, the ions of sodium chloride: Na^+, Cl^-.)

Magnesium

Magnesium is a chemical element with the symbol Mg, atomic number 12, atomic weight 24.3050 and common oxidation number +2.

Magnesium, an alkaline earth metal, is the ninth most abundant element in the universe by mass. The commonness of Magnesium is related to the fact that it is easily built up in supernova stars from a sequential addition of three helium nuclei to carbon.

Metathesis

Metathesis is a biomolecular process involving the exchange of bonds between the two reacting chemical species, which results in the creation of products with similar or identical bonding affiliations. To illustrate, consider two chemical species, AB and CD, which react to give AD and CB:

$$AB + CD \rightarrow AD + CB$$

These chemical species can either be ionic or covalent. When referring to precipitation reactions between solutions of ions in inorganic chemistry, these were formerly referred to as double displacement or double replacement reactions, though these terms are still encouraged.

Silver

Silver is a chemical element with the chemical symbol Ag and atomic number 47. A soft, white, lustrous transition metal, it has the highest electrical conductivity of any element and the highest thermal conductivity of any metal. The metal occurs naturally in its pure, free form, as an alloy with gold (electrum) and other metals, and in minerals such as argentite and chlorargyrite.

Chapter 4. Chemical Reactions

Chapter 4. Chemical Reactions

Boiling	Boiling, a type of phase transition, is the rapid vaporization of a liquid, which typically occurs when a liquid is heated to its Boiling point, the temperature at which the vapor pressure of the liquid is equal to the pressure exerted on the liquid by the surrounding environmental pressure. Thus, a liquid may also boil when the pressure of the surrounding atmosphere is sufficiently reduced, such as the use of a vacuum pump or at high altitudes. Boiling occurs in three characteristic stages, which are nucleate, transition and film Boiling.
Copper	Copper is a chemical element with the symbol Cu and atomic number 29. It is a ductile metal with very high thermal and electrical conductivity. Pure Copper is rather soft and malleable and a freshly-exposed surface has a pinkish or peachy color.
Melting	Melting is a physical process that results in the phase change of a substance from a solid to a liquid. The internal energy of a solid substance is increased, typically by the application of heat or pressure, resulting in a rise of its temperature to the Melting point, at which the rigid ordering of molecular entities in the solid breaks down to a less-ordered state and the solid liquefies. An object that has melted completely is molten.
Iron	Iron is a chemical element with the symbol Fe and atomic number 26. Iron is a group 8 and period 4 element. Iron and Iron alloys (steels) are by far the most common metals and the most common ferromagnetic materials in everyday use.
Sodium nitrate	Sodium nitrate is the chemical compound with the formula $NaNO_3$. This salt, also known as "Chile saltpeter" or "Peru saltpeter" (to distinguish it from ordinary saltpeter, potassium nitrate), is a white solid which is very soluble in water. The mineral form is also known as nitratine or soda niter.
Freezing	In physical science, Freezing or solidification is the process in which a liquid turns into a solid when cold enough. The Freezing point is the temperature at which this happens. Melting, the process of turning a solid to a liquid, is almost the exact opposite of Freezing.
Citric acid	Citric acid is a weak organic acid, and it is a natural preservative and is also used to add an acidic taste to foods and soft drinks. In biochemistry, it is important as an intermediate in the Citric acid cycle and therefore occurs in the metabolism of virtually all living things. It can also be used as an environmentally benign cleaning agent.
Litmus	Litmus is a water-soluble mixture of different dyes extracted from lichens, especially Roccella tinctoria. The mixture has CAS number 1393-92-6. It is often absorbed onto filter paper.
Nitric acid	Nitric acid, also known as aqua fortis and spirit of nitre, is a highly corrosive and toxic strong acid that can cause severe burns. Colorless when pure, older samples tend to acquire a yellow cast due to the accumulation of oxides of nitrogen. If the solution contains more than 86% Nitric acid, it is referred to as fuming Nitric acid.

Chapter 4. Chemical Reactions

Chapter 4. Chemical Reactions

Hydronium	In chemistry, Hydronium is the common name for the aqueous cation H_3O^+, the simplest type of oxonium ion, produced by protonation of water. It is the positive ion present when an Arrhenius acid is dissolved in water, as Arrhenius acid molecules in solution give up a proton (a positive hydrogen ion, H^+) to the surrounding water molecules (H_2O.) It is the presence of Hydronium ion relative to hydroxide that determines a solution"s pH. The molecules in pure water auto-dissociate into Hydronium and hydroxide ions in the following equillibrium: $$2H_2O \; OH^- + H_3O^+$$ In pure water, there is an equal number of hydroxide and Hydronium ions, and the pH is perfectly neutral (7.0.)
Hydrofluoric acid	Hydrofluoric acid is a solution of hydrogen fluoride in water. While it is extremely corrosive and dangerous to handle, it is technically a weak acid. Hydrogen fluoride, often in the aqueous form as Hydrofluoric acid, is a valued source of fluorine, being the precursor to numerous pharmaceuticals (e.g., Prozac), diverse polymers (e.g., Teflon), and most other synthetic materials that contain fluorine.
Strong acid	A Strong acid is an acid that dissociates completely in an aqueous solution (not in the case of sulfuric acid as it is diprotic), or in other terms, with a $pK_a < -1.74$. This generally means that in aqueous solution at standard temperature and pressure, the concentration of hydronium ions is equal to the concentration of Strong acid introduced to the solution. While Strong acid s are generally assumed to be the most corrosive, this is not always true.
Hydrobromic acid	Hydrobromic acid is a strong acid formed by dissolving the diatomic molecule hydrogen bromide in water. "Constant boiling" Hydrobromic acid is an aqueous solution that distills at 124.3 °C and contains 47.6% HBr by weight. Hydrobromic acid has a pK_a of −9, making it a stronger acid than hydrochloric acid, but not as strong as hydroiodic acid.
Lithium	Lithium is the chemical element with atomic number 3, and is represented by the symbol Li. It is a soft alkali metal with a silver-white color. Under standard conditions it is the lightest metal and the least dense solid element.
Lithium hydroxide	Lithium hydroxide is a corrosive alkali hydroxide. It is a white hygroscopic crystalline material. It is soluble in water, and slightly soluble in ethanol.
Neutralization	In chemistry, Neutralization, or neutralisation , is a chemical reaction in which an acid and a base or alkali (soluble base) react to produce salt and water (H_2O.) During the process, hydrogen ions H^+ (a bare proton) from the acid (proton donor) or a hydronium ion H_3O^+ and hydroxide ions OH^- or oxide ions O^{2-} from the base (proton acceptor) react together to form a water molecule H_2O. In the process, a salt is also formed when the anion from acid and the cation from base react together.

Chapter 4. Chemical Reactions

Chapter 4. Chemical Reactions

	Neutralization reactions are generally classified as exothermic since heat is released into the surroundings.
Nitrogen	Nitrogen is a chemical element that has the symbol N and atomic number 7 and atomic mass 14.00674 u. Elemental Nitrogen is a colorless, odorless, tasteless and mostly inert diatomic gas at standard conditions, constituting 78% by volume of Earth"s atmosphere. Many industrially important compounds, such as ammonia, nitric acid, organic nitrates , and cyanides, contain Nitrogen.
Salt	Salt is a dietary mineral composed primarily of sodium chloride that is essential for animal life, but toxic to most land plants. Salt flavor is one of the basic tastes, an important preservative and a popular food seasoning.
Complex	In chemistry, a Complex is a structure consisting of a central atom or molecule connected to surrounding atoms or molecules. Originally, a Complex implied a reversible association of molecules, atoms, or ions through weak chemical bonds. As applied to coordination chemistry, this meaning has evolved.
Potassium hydroxide	Potassium hydroxide is the inorganic compound with the formula KOH. Along with sodium hydroxide, this colourless solid is a prototypical "strong base". It has many industrial and niche applications. Most applications exploit its reactivity toward acids and its corrosive nature.
Nitrous acid	Nitrous acid is a weak and monobasic acid known only in solution and in the form of nitrite salts. Nitrous acid is used to make diazides from amines; this occurs by nucleophilic attack of the amine onto the nitrite, reprotonation by the surrounding solvent, and double-elimination of water. The diazide can then be liberated as a carbene.
Sulfuric acid	Sulfuric (or sulphuric) acid, H_2SO_4, is a strong mineral acid. It is soluble in water at all concentrations. Sulfuric acid has many applications, and is one of the top products of the chemical industry.
Acid salts	Acid salts are a class of salts formed when a dibasic or tribasic acid has been neutralized to some degree. Because the acid is only partially neutralized, one or more replaceable protons remain. Typically this will lead to a formula with one or more metal ions, one or more protons, and an anion, such as sodium bicarbonate ($NaHCO_3$), sodium hydrosulfide (NaHS), sodium bisulfate ($NaHSO_4$), monosodium phosphate (NaH_2PO_4), and disodium phosphate (Na_2HPO_4.)
Carbonic acid	Carbonic acid (ancient name acid of air or aerial acid) has the formula H_2CO_3. It is also a name sometimes given to solutions of carbon dioxide in water, which contain small amounts of H_2CO_3. The salts of Carbonic acid s are called bicarbonates (or hydrogen carbonates) and carbonates.

Chapter 4. Chemical Reactions

Chapter 4. Chemical Reactions

Hydrogen sulfide	Hydrogen sulfide is the chemical compound with the formula H_2S. This colorless, toxic and flammable gas is partially responsible for the foul odor of rotten eggs and flatulence. It often results from the bacterial break down of sulfites in nonorganic matter in the absence of oxygen, such as in swamps and sewers (anaerobic digestion.) It also occurs in volcanic gases, natural gas and some well waters.
Sulfur	Sulfur or sulphur is the chemical element that has the atomic number 16. It is denoted with the symbol S. It is an abundant, multivalent non-metal. Sulfur in its native form is a yellow crystalline solid.
Ionization	Ionization is the physical process of converting an atom or molecule into an ion by adding or removing charged particles such as electrons or other ions. This is often confused with dissociation (chemistry.) The process works slightly differently depending on whether an ion with a positive or a negative electric charge is being produced.
Zinc	Zinc is a metallic chemical element with the symbol Zn and atomic number 30. It is a first-row transition metal in group 12 of the periodic table. Zinc is chemically similar to magnesium because its ion is of similar size and its only common oxidation state is +2.
Atomic weight	Atomic weight (symbol: A_r) is a dimensionless physical quantity, the ratio of the average mass of atoms of an element to 1/12 of the mass of an atom of carbon-12. The term is usually used, without further qualification, to refer to the standard Atomic weight s published at regular intervals by the International Union of Pure and Applied Chemistry (IUPAC) and which are intended to be applicable to normal laboratory materials
Oxidation number	The Oxidation number of a central atom in a coordination compound is the charge that it would have if all the ligands were removed along with the electron pairs that were shared with the central atom. It is used in the nomenclature of inorganic compounds. It is represented by a Roman numeral; the plus sign is omitted for positive Oxidation number s.
Oxygen	Oxygen and -γενÎ®ς (-genÄ"s) (producer, literally begetter) is the element with atomic number 8 and represented by the symbol O. It is a member of the chalcogen group on the periodic table, and is a highly reactive nonmetallic period 2 element that readily forms compounds (notably oxides) with almost all other elements. At standard temperature and pressure two atoms of the element bind to form di Oxygen, a colorless, odorless, tasteless diatomic gas with the formula O_2. Oxygen is the third most abundant element in the universe by mass after hydrogen and helium and the most abundant element by mass in the Earth"s crust.
Redox	Redox describes all chemical reactions in which atoms have their oxidation number (oxidation state) changed. This can be either a simple Redox process such as the oxidation of carbon to yield carbon dioxide or the reduction of carbon by hydrogen to yield methane (CH_4), or it can be a complex process such as the oxidation of sugar in the human body through a series of very complex electron transfer processes.

Chapter 4. Chemical Reactions

Chapter 4. Chemical Reactions

	The term Redox comes from the two concepts of reduction and oxidation.
Chlorin	In organic chemistry, a Chlorin is a large heterocyclic aromatic ring consisting, at the core, of three pyrroles and one pyrroline coupled through four methine linkages. Unlike a porphyrin, a Chlorin is therefore largely aromatic but not aromatic through the entire circumference of the ring. Magnesium-containing Chlorin s are called chlorophylls, and are the central photosensitive pigment in chloroplasts.
Chlorine	Chlorine . As the chloride ion, which is part of common salt and other compounds, it is abundant in nature and necessary to most forms of life, including humans. In its elemental form (Cl_2 or "di Chlorine ") under standard conditions, Chlorine is a powerful oxidant and is used in bleaching and disinfectants.
Polyaniline	Polyaniline is a conducting polymer of the semi-flexible rod polymer family. Although it was discovered over 150 years ago, only recently has Polyaniline captured the attention of the scientific community due to the discovery of its high electrical conductivity. Amongst the family of conducting polymers, Polyaniline is unique due to its ease of synthesis, environmental stability, and simple doping/dedoping chemistry.
Polyatomic ion	A Polyatomic ion is a charged species composed of two or more atoms covalently bonded or of a metal complex that can be considered as acting as a single unit in the context of acid and base chemistry or in the formation of salts. The prefix poly- means many in Greek, but even ions of two atoms are commonly referred to as polyatomic. In older literature, a Polyatomic ion is also referred to as a radical, and less commonly, as a radical group.
Reducing agent	A Reducing agent is the element or compound in a redox reaction that reduces another species. In doing so, it becomes oxidized, and is therefore the electron donor in the redox. For example consider the following reaction: $[Fe_6]^{4-} + 1/2\ Cl_2 \rightarrow [Fe_6]^{3-} + Cl^-$ The Reducing agent in this reaction is ferrocyanide: it donates an electron, converting to ferricyanide, simultaneous with the reduction of chlorine to chloride.
Combination reaction	A Combination reaction or a Synthesis Reaction is a general category of a chemical reaction (the term usually refers to an inorganic chemical reaction), in which two or more reagents are chemically bonded together to produce a single product.when two or more substances combine to form a single product, it is known as a Combination reaction.
	A Combination reaction can be of three types:a) Between 2 elementsb) Between 2 compoundsc) Between an element and a compound Examples:a) $2Ag + O_2 \rightarrow 2AgO$ b) $O + H_2O \rightarrow H_2O_2$ c) $2C(s) + 2O_2(g) \rightarrow 2CO_2\ (g)$ d) $CaO + H_2O \rightarrow Ca(OH)_2$

Chapter 4. Chemical Reactions

Chapter 4. Chemical Reactions

Antimony	Antimony) is a chemical element with the symbol Sb and atomic number 51. A metalloid, Antimony has four allotropic forms. The stable form of Antimony is a blue-white metalloid.
Manganese	Manganese is a chemical element, designated by the symbol Mn. It has the atomic number 25. It is found as a free element in nature, and in many minerals.
Mercury oxide	Mercury oxide has a formula of HgO and a formula weight of 216.6. It has a red or orange color. Mercury oxide is a solid at room temperature and pressure.
Single-displacement reaction	A single-displacement reaction is a type of oxidation-reduction chemical reaction when an element or ion moves out of one compound and into another. This is usually written as 　　A + BX → AX + B This will occur if A is more reactive than B. You can refer to the reactivity series to be sure of this.
Combustion	Combustion or burning is a complex sequence of exothermic chemical reactions between a fuel (usually a hydrocarbon) and an oxidant accompanied by the production of heat or both heat and light in the form of either a glow or flames, appearance of light flickering. Direct Combustion by atmospheric oxygen is a reaction mediated by radical intermediates. The conditions for radical production are naturally produced by thermal runaway, where the heat generated by Combustion is necessary to maintain the high temperature necessary for radical production.
Molar concentration	In chemistry, molar concentration is a measure of the concentration of a solute in a solution ionic in thermodynamics the use of molar concentration is often not very convenient, because the volume of most solutions slightly depends on temperature due to thermal expansion. This problem is usually resolved by introducing temperature correction factors, or by using a temperature-independent measure of concentration such as molality.
Solvent	A Solvent is a liquid or gas that dissolves a solid, liquid resulting in a solution. The most common Solvent in everyday life is water. Most other commonly-used Solvent s are organic (carbon-containing) chemicals.
Atom	The Atom is a basic unit of matter consisting of a dense, central nucleus surrounded by a cloud of negatively charged electrons. The atomic nucleus contains a mix of positively charged protons and electrically neutral neutrons (except in the case of hydrogen-1, which is the only stable nuclide with no neutron.) The electrons of an Atom are bound to the nucleus by the electromagnetic force.

Chapter 4. Chemical Reactions

Chapter 4. Chemical Reactions

Molecule	A Molecule is defined as a sufficiently stable, electrically neutral group of at least two atoms in a definite arrangement held together by very strong (covalent) chemical bonds. Molecule s are distinguished from polyatomic ions in this strict sense. In organic chemistry and biochemistry, the term Molecule is used less strictly and also is applied to charged organic Molecule s and bio Molecule s.
Mole	The mole is a unit of amount of substance: it is an SI base unit, and one of the few units used to measure this physical quantity. The name "mole" was coined in German by Wilhelm Ostwald in 1893, although the related concept of equivalent mass had been in use at least a century earlier. The name is assumed to be derived from the word Molekül (molecule.)
Moles	Moles are the majority of the members of the mammal family Talpidae in the order Soricomorpha. Although most moles burrow, some species are aquatic or semi-aquatic. moles have cylindrical bodies covered in fur, with small or covered eyes; the ears are generally not visible.
Analytical chemistry	Analytical chemistry is the study of the chemical composition of natural and artificial materials. Properties studied in Analytical chemistry include geometric features such as molecular morphologies and distributions of species, as well as features such as composition and species identity. Unlike the sub disciplines inorganic chemistry and organic chemistry, Analytical chemistry is not restricted to any particular type of chemical compound or reaction.The contributions made by analytical chemists have played critical roles in the sciences ranging from the development of concepts and theories (pure science) to a variety of practical applications, such as biomedical applications, environmental monitoring, quality control of industrial manufacturing and forensic science (applied science.)
Chemistry	In the history of science, the etymology of the word Chemistry is a debatable issue. It is agreed that the word "alchemy" is a European one, derived from Arabic, but the origin of the root word, chem, is uncertain. Words similar to it have been found in most ancient languages, with different meanings, but conceivably somehow related to alchemy.
Enthalpy	In thermodynamics and molecular chemistry, the Enthalpy is a thermodynamic property of a fluid. It can be used to calculate the heat transfer during a quasistatic process taking place in a closed thermodynamic system under constant pressure. Enthalpy H is an arbitrary concept but the Enthalpy change ΔH is more useful because it is equal to the change in the internal energy of the system, plus the work that the system has done on its surroundings.
Nickel	Nickel is a chemical element, with the chemical symbol Ni and atomic number 28. It is a silvery-white lustrous metal with a slight golden tinge. It is one of the four ferromagnetic elements at about room temperature.

Chapter 5. The Gaseous State

Mercury	There are seven isotopes of mercury with Hg-202 being the most abundant (29.86%.) The longest-lived radioisotopes are ^{194}Hg with a half-life of 444 years, and ^{203}Hg with a half-life of 46.612 days. Most of the remaining radioisotopes have half-lives that are less than a day.
Mole	The mole is a unit of amount of substance: it is an SI base unit, and one of the few units used to measure this physical quantity. The name "mole" was coined in German by Wilhelm Ostwald in 1893, although the related concept of equivalent mass had been in use at least a century earlier. The name is assumed to be derived from the word Molekül (molecule.)
Moles	Moles are the majority of the members of the mammal family Talpidae in the order Soricomorpha. Although most moles burrow, some species are aquatic or semi-aquatic. moles have cylindrical bodies covered in fur, with small or covered eyes; the ears are generally not visible.
Robert Boyle	Robert Boyle was a natural philosopher, chemist, physicist, inventor, and gentleman scientist, also noted for his writings in theology. He is best known for the formulation of Boyle"s law. Although his research and personal philosophy clearly has its roots in the alchemical tradition, he is largely regarded today as the first modern chemist, and therefore one of the founders of modern chemistry.
Molar concentration	In chemistry, molar concentration is a measure of the concentration of a solute in a solution ionic in thermodynamics the use of molar concentration is often not very convenient, because the volume of most solutions slightly depends on temperature due to thermal expansion. This problem is usually resolved by introducing temperature correction factors, or by using a temperature-independent measure of concentration such as molality.
Dalton	Dalton is an ab initio quantum chemistry software program. It is capable of calculating various molecular properties using the Hartree-Fock, MP2, MCSCF and coupled cluster theories. Version 2.0 of Dalton added support for density functional theory calculations.
Copper	Copper is a chemical element with the symbol Cu and atomic number 29. It is a ductile metal with very high thermal and electrical conductivity. Pure Copper is rather soft and malleable and a freshly-exposed surface has a pinkish or peachy color.
Dumas method	The Dumas method in analytical chemistry is a method for the quantitative determination of nitrogen in chemical substances based on a method first described by Jean-Baptiste Dumas over a century and a half ago. An automated instrumental technique has been developed which is capable of rapidly measuring the crude protein concentration of food samples and is beginning to compete with the Kjeldahl method as the standard method of analysis for protein content for some foodstuffs. The method consists of combusting a sample of known mass in a high temperature (about 900 °C) chamber in the presence of oxygen.

Chapter 5. The Gaseous State

Chapter 5. The Gaseous State

Complex	In chemistry, a Complex is a structure consisting of a central atom or molecule connected to surrounding atoms or molecules. Originally, a Complex implied a reversible association of molecules, atoms, or ions through weak chemical bonds. As applied to coordination chemistry, this meaning has evolved.
Ion	An Ion is an atom or molecule where the total number of electrons is not equal to the total number of protons, giving it a net positive or negative electrical charge. Since protons are positively charged and electrons are negatively charged, if there are more electrons than protons, the atom or molecule will be negatively charged. This is called an an Ion , from the Greek á¼€vÎ¬ , meaning "up".
Cadmium	Cadmium is a chemical element with the symbol Cd and atomic number 48. The soft, bluish-white transition metal is chemically similar to the two other metals in group 12 , zinc and mercury. Similar to zinc it prefers oxidation state +2 in most of its compounds and similar to mercury it shows a low melting point for a transition metal.
Amount of substance	In physical sciences, the Amount of substance, n, of a sample can be defined informally as the number of some specified elementary entities (usually either atoms or ions but where this number is expressed in terms of some standard batch size. This is analogous to expressing the number of years in terms of centuries see below for a formal definition. What makes amount-of-substance a useful quantity--even though it may seem that it is a redundant one, given that it seems to contain the same information as does the total number of entities--is its particular convenience in real-world experimental settings, a convenience that largely stems from the specific choice that was made for the definition of the standard batch size.
Nitrogen	Nitrogen is a chemical element that has the symbol N and atomic number 7 and atomic mass 14.00674 u. Elemental Nitrogen is a colorless, odorless, tasteless and mostly inert diatomic gas at standard conditions, constituting 78% by volume of Earth"s atmosphere. Many industrially important compounds, such as ammonia, nitric acid, organic nitrates , and cyanides, contain Nitrogen.
Polymer	A Polymer is a large molecule composed of repeating structural units typically connected by covalent chemical bonds. While Polymer in popular usage suggests plastic, the term actually refers to a large class of natural and synthetic materials with a variety of properties. Due to the extraordinary range of properties accessible in Polymer ic materials , they have come to play an essential and ubiquitous role in everyday life - from plastics and elastomers on the one hand to natural bio Polymer s such as DNA and proteins that are essential for life on the other.
Concentration	In chemistry, Concentration is the measure of how much of a given substance there is mixed with another substance. This can apply to any sort of chemical mixture, but most frequently the concept is limited to homogeneous solutions, where it refers to the amount of solute in the solvent.

Chapter 5. The Gaseous State

Chapter 5. The Gaseous State

	To concentrate a solution, one must add more solute, or reduce the amount of solvent (for instance, by selective evaporation.)
Gas constant	The Gas constant is a physical constant which is featured in a large number of fundamental equations in the physical sciences, such as the ideal gas law and the Nernst equation. It is equivalent to the Boltzmann constant, but expressed in units of energy (i.e. the pressure-volume product) per kelvin per mole (rather than energy per kelvin per particle.) Its value is $$R = 8.314\,472(15)\,\frac{\mathrm{J}}{\mathrm{K\,mol}}.$$ The two digits in parentheses are the uncertainty (standard deviation) in the last two digits of the value.
Molecular mass	The Molecular mass of a substance, frequently referred by the older term molecular weight and abbreviated as MW, is the mass of one molecule of that substance, relative to the unified atomic mass unit u (equal to 1/12 the mass of one isotope of carbon-12.) This is distinct from the relative Molecular mass of a molecule, which is the ratio of the mass of that molecule to 1/12 of the mass of carbon 12 and is a dimensionless number. Relative Molecular mass is abbreviated to M_r.
Oxygen	Oxygen and -γενî®ς (-genÄ"s) (producer, literally begetter) is the element with atomic number 8 and represented by the symbol O. It is a member of the chalcogen group on the periodic table, and is a highly reactive nonmetallic period 2 element that readily forms compounds (notably oxides) with almost all other elements. At standard temperature and pressure two atoms of the element bind to form di Oxygen , a colorless, odorless, tasteless diatomic gas with the formula O_2. Oxygen is the third most abundant element in the universe by mass after hydrogen and helium and the most abundant element by mass in the Earth"s crust.
Bromine	Bromine , Greek: βρά¿¶μος, brómos, meaning "stench "), is a chemical element with the symbol Br and atomic number 35. A halogen element, Bromine is a reddish-brown volatile liquid at standard room temperature that is intermediate in reactivity between chlorine and iodine. Bromine vapours are corrosive and toxic.
Group	In chemistry, a Group is a vertical column in the periodic table of the chemical elements. The name family is derived from the fact that the elements share similar characteristics and traits, just as members of any human family would. There are 18 groups in the standard periodic table.
Vapor	A Vapor or vapour is a substance in the gas phase at a temperature lower than its critical temperature. This means that the Vapor can be condensed to a liquid or to a solid by increasing its pressure, without reducing the temperature.

Chapter 5. The Gaseous State

Chapter 5. The Gaseous State

	For example, water has a critical temperature of 374°C (or 647 K) which is the highest temperature at which liquid water can exist.
Iron	Iron is a chemical element with the symbol Fe and atomic number 26. Iron is a group 8 and period 4 element. Iron and Iron alloys (steels) are by far the most common metals and the most common ferromagnetic materials in everyday use.
Sodium	Sodium is a metallic element with a symbol Na and atomic number 11. It is a soft, silvery-white, highly reactive metal and is a member of the alkali metals within "group 1" (formerly known as "group IA".) It has only one stable isotope, ^{23}Na.
Stoichiometry	Stoichiometry is the calculation of quantitative (measurable) relationships of the reactants and products in a balanced chemical reaction (chemicals.) It can be used to calculate quantities such as the amount of products that can be produced with the given reactants and percent yield. "Stoichiometry" is derived from the Greek words στοιχεά¿–ov and μι̂τρον (metron, meaning measure.)
Carbon	Carbon is the chemical element with symbol C and atomic number 6. As a member of group 14 on the periodic table, it is nonmetallic and tetravalent--making four electrons available to form covalent chemical bonds. There are three naturally occurring isotopes, with ^{12}C and ^{13}C being stable, while ^{14}C is radioactive, decaying with a half-life of about 5730 years.
Solution	In chemistry, a Solution is a homogeneous mixture composed of two or more substances. In such a mixture, a solute is dissolved in another substance, known as a solvent. Gases may dissolve in liquids, for example, carbon dioxide or oxygen in water.
Helium	Helium is the chemical element with atomic number 2, and is represented by the symbol He. It is a colorless, odorless, tasteless, non-toxic, inert monatomic gas that heads the noble gas group in the periodic table. Its boiling and melting points are the lowest among the elements and it exists only as a gas except in extreme conditions.
Hydrogen	Hydrogen is the chemical element with atomic number 1. It is represented by the symbol H. At standard temperature and pressure, Hydrogen is a colorless, odorless, nonmetallic, tasteless, highly flammable diatomic gas with the molecular formula H_2. With an atomic weight of 1.007 94 u, Hydrogen is the lightest element.
Crystal	A Crystal or Crystal line solid is a solid material whose constituent atoms, molecules, or ions are arranged in an orderly repeating pattern extending in all three spatial dimensions. The scientific study of Crystal s and Crystal formation is Crystal lography. The process of Crystal formation via mechanisms of Crystal growth is called Crystal lization or solidification.

Chapter 5. The Gaseous State

Chapter 5. The Gaseous State

Crystal structure	In mineralogy and crystallography, a Crystal structure is a unique arrangement of atoms in a crystal. A Crystal structure is composed of a motif, a set of atoms arranged in a particular way, and a lattice. Motifs are located upon the points of a lattice, which is an array of points repeating periodically in three dimensions.
Mole fraction	In chemistry, mole fraction x (also, and more correctly, known as the amount fraction) is a way of expressing the composition of a mixture. The mole fraction of each component i is defined as its amount of substance n_i divided by the total amount of substance in the system, n $$x_i \stackrel{\text{def}}{=} \frac{n_i}{n}$$ where $$n = \sum_i n_i$$ The sum is over all components, including the solvent in the case of a chemical solution. As an example, if a mixture is obtained by dissolving 10 moles of sucrose in 90 moles of water, the mole fraction of sucrose in that mixture is 0.1.
Fraction	A Fraction in chemistry is a quantity collected from a sample or batch of a substance in a fractionating separation process. In such a process, a mixture is separated into fractions, which have compositions that vary according to a gradient. A Fraction can be defined as a group of chemicals that have similar boiling points.
Fractional distillation	Fractional distillation is the separation of a mixture into its component parts such as in separating chemical compounds by their boiling point by heating them to a temperature at which several fractions of the compound will evaporate. It is a special type of distillation. Generally the component parts boil at less than 25 °C from each other under a pressure of one atmosphere (atm.)
Vapor pressure	Vapor pressure, is the pressure of a vapor in equilibrium with its non-vapor phases. All liquids and solids have a tendency to evaporate to a gaseous form, and all gases have a tendency to condense back into their original form At any given temperature, for a particular substance, there is a pressure at which the gas of that substance is in dynamic equilibrium with its liquid or solid forms.
Water	Water is the most abundant molecule on Earth"s surface, constituting about 75% of the Earth"s surface. In nature it exists in liquid, solid, and gaseous states. It is in dynamic equilibrium between the liquid and gas states at standard temperature and pressure.
Hydrochloric acid	Hydrochloric acid is the solution of hydrogen chloride (HCl) in water. It is a highly corrosive, strong mineral acid and has major industrial uses. It is found naturally in gastric acid.

Chapter 5. The Gaseous State

Chapter 5. The Gaseous State

Zinc	Zinc is a metallic chemical element with the symbol Zn and atomic number 30. It is a first-row transition metal in group 12 of the periodic table. Zinc is chemically similar to magnesium because its ion is of similar size and its only common oxidation state is +2.
Acid	An Acid is traditionally considered any chemical compound that, when dissolved in water, gives a solution with a hydrogen ion activity greater than in pure water, i.e. a pH less than 7.0. That approximates the modern definition of Johannes Nicolaus Brønsted and Martin Lowry, who independently defined an Acid as a compound which donates a hydrogen ion (H^+) to another compound (called a base.) Common examples include acetic Acid and sulfuric Acid (used in car batteries.)
Atomic theory	In chemistry and physics, Atomic theory is a theory of the nature of matter, which states that matter is composed of discrete units called atoms, as opposed to the obsolete notion that matter could be divided into any arbitrarily small quantity. It began as a philosophical concept in ancient Greece and India and entered the scientific mainstream in the early 19th century when discoveries in the field of chemistry showed that matter did indeed behave as if it were made up of particles. The word "atom" was applied to the basic particle that constituted a chemical element, because the chemists of the era believed that these were the fundamental particles of matter.
Intermolecular forces	In physics, chemistry, and biology, Intermolecular forces are forces that act between stable molecules or between functional groups of macromolecules. Intermolecular forces include momentary attractions between molecules, diatomic free elements, and individual atoms. They differ from covalent and ionic bonding in that they are not stable, but are caused by momentary polarization of particles.
Sulfur	Sulfur or sulphur is the chemical element that has the atomic number 16. It is denoted with the symbol S. It is an abundant, multivalent non-metal. Sulfur in its native form is a yellow crystalline solid.
Nuclear fission	In nuclear physics and nuclear chemistry, Nuclear fission is a nuclear reaction in which the nucleus of an atom splits into smaller parts, often producing free neutrons and lighter nuclei, which may eventually produce photons (in the form of gamma rays.) Fission of heavy elements is an exothermic reaction which can release large amounts of energy both as electromagnetic radiation and as kinetic energy of the fragments (heating the bulk material where fission takes place.) For fission to produce energy, the total binding energy of the resulting elements has to be higher than that of the starting element.
Uranium	Uranium is a silvery-white metallic chemical element in the actinide series of the periodic table that has the symbol U and atomic number 92. Besides its 92 protons, a Uranium nucleus can have between 141 and 146 neutrons, with 146 and 143 (U-235) in its most common isotopes. The number of electrons in a Uranium atom is 92, 6 of them valence electrons.

Chapter 5. The Gaseous State

Chapter 5. The Gaseous State

Background radiation	Background radiation is constantly present in the environment and is emitted from a variety of natural and artificial sources. Primary contributions come from: • Sources in the earth. These include sources in food and water, which are incorporated in the body, and in building materials and other products that incorporate those radioactive sources; • Sources from space, in the form of cosmic rays; • Sources in the atmosphere. One significant contribution comes from the radon gas that is released from the Earth"s crust and subsequently decays into radioactive atoms that become attached to airborne dust and particulates. Another contribution arises from the radioactive atoms produced in the bombardment of atoms in the upper atmosphere by high-energy cosmic rays. About 3% of Background radiation comes from other man-made sources such as: • Self-luminous dials and signs • Global radioactive contamination due to historical nuclear weapons testing • Nuclear power station or nuclear fuel reprocessing accidents (though these are rare) • Normal operation of facilities used for nuclear power and scientific research • Emissions from burning fossil fuels, such as coal fired power plants • Emissions from nuclear medicine facilities and patients • Emissions from the improper disposal or recycling of radioactive materials used in nuclear medicine Accidental exposure to man-made radioactive substances can result in radiation exposure that is many times that received from background sources, whether natural or man-made. Additionally, radiation therapy can cause relatively high levels of exposure. However, when it comes to Background radiation, naturally occurring sources are responsible for the vast majority of radiation exposure.
Vanadium	Vanadium is the chemical element with the symbol V and atomic number 23. It is a soft, silvery grey, ductile transition metal. The formation of an oxide layer stabilizes the metal against oxidation.
Covalent radius	The Covalent radius, r_{cov}, is a measure of the size of atom that forms part of a covalent bond. It is measured either in picometres (pm) or ångströms (Å), with 1 Å = 100 pm. In principle, the sum of the two covalent radii should equal the covalent bond length between two atoms.
Acid rain	Acid rain is rain or any other form of precipitation that is unusually acidic. It has harmful effects on plants, aquatic animals, and infrastructure. Acid rain is mostly caused by human emissions of sulfur and nitrogen compounds which react in the atmosphere to produce acids.

Chapter 5. The Gaseous State

Ozone	Ozone or trioxygen (O_3) is a triatomic molecule, consisting of three oxygen atoms. It is an allotrope of oxygen that is much less stable than the diatomic O_2. Ground-level Ozone is an air pollutant with harmful effects on the respiratory systems of animals.

Chapter 5. The Gaseous State

Chapter 6. Thermochemistry

Barium	Barium is a chemical element. It has the symbol Ba, and atomic number 56. Barium is a soft silvery metallic alkaline earth metal.
Thermochemistry	In thermodynamics and physical chemistry, Thermochemistry is the study of the energy evolved or absorbed in chemical reactions and any physical transformations, such as melting and boiling. Thermochemistry, generally, is concerned with the energy exchange accompanying transformations, such as mixing, phase transitions, chemical reactions, and including calculations of such quantities as the heat capacity, heat of combustion, heat of formation, enthalpy, and free energy.

Thermochemistry rests on two generalizations:

1. Lavoisier and Laplace"s law (1782): the heat exchange accompanying a transformation is equal and opposite to the heat exchange accompanying the reverse transformation.
2. Hess"s law (1840): the heat exchange accompanying a transformation is the same whether the process occurs in one or both steps

Both of these statements preceded the first law of thermodynamics (1850) and helped in its formulation.

Lavoisier, Laplace and Hess also investigated specific heat and latent heat, although it was Joseph Black who made the most important contributions to the development of latent energy changes.

Entropy	In thermodynamics, Entropy is often associated with the amount of order, disorder, and/or chaos in a thermodynamic system. This stems from Rudolf Clausius" 1862 assertion that any thermodynamic processes always "admits to being reduced to the alteration in some way or another of the arrangement of the constituent parts of the working body" and that internal work associated with these alterations is quantified energetically by a measure of "Entropy" change, according to the following differential expression:

$$\int \frac{\delta Q}{T} \geq 0$$

In the years to follow, Ludwig Boltzmann translated these "alterations" into that of a probabilistic view of order and disorder in gas phase molecular systems.

In recent years, in some chemistry publications, there has been a shift away from using the terms "order" and "disorder" to that of the concept of energy dispersion to describe Entropy, among other theories.

Hydroxide	In chemistry, Hydroxide is the name for the diatomic anion OH⁻, consisting of oxygen and hydrogen atoms, usually derived from the dissociation of a base. It is one of the simplest diatomic ions known.

Inorganic compounds that contain the hydroxyl group are referred to as Hydroxide s.

Chapter 6. Thermochemistry

Chapter 6. Thermochemistry

Internal energy	In thermodynamics, the Internal energy of a thermodynamic system denoted by U is the total of the kinetic energy due to the motion of molecules (translational, rotational, vibrational) and the potential energy associated with the vibrational and electric energy of atoms within molecules or crystals. It includes the energy in all of the chemical bonds, and the energy of the free, conduction electrons in metals. One can also calculate the Internal energy of electromagnetic or blackbody radiation.
Emission	In physics, Emission is the process by which the energy of a photon is released by another entity, for example, by an atom whose electrons make a transition between two electronic energy levels. The emitted energy is in the form of a photon. The emittance of an object quantifies how much light is emitted by it.
Endothermic	In thermodynamics, the word Endothermic "within-heating" describes a process or reaction that absorbs energy in the form of heat. Its etymology stems from the Greek prefix endo-, meaning "inside" and the Greek suffix -thermic, meaning "to heat". The opposite of an Endothermic process is an exothermic process, one that releases energy in the form of heat.
Oxygen	Oxygen and -γενÎ®ς (-genÄ"s) (producer, literally begetter) is the element with atomic number 8 and represented by the symbol O. It is a member of the chalcogen group on the periodic table, and is a highly reactive nonmetallic period 2 element that readily forms compounds (notably oxides) with almost all other elements. At standard temperature and pressure two atoms of the element bind to form di Oxygen , a colorless, odorless, tasteless diatomic gas with the formula O_2. Oxygen is the third most abundant element in the universe by mass after hydrogen and helium and the most abundant element by mass in the Earth"s crust.
Methanation	Methanation is a physical-chemical process to generate Methane from a mixture of various gases out of biomass fermentation or thermo-chemical gasification. The main components are carbon monoxide and hydrogen. The following main process describes the Methanation: $$CO + 3\,H_2 \rightarrow CH_4 + H_2O$$ This process is used for the generation of biogenous natural gas substitute, which can be fed into the gas grid. Methanation is the reverse reaction of steam methane reforming, which converts methane into synthesis gas.
Enthalpy	In thermodynamics and molecular chemistry, the Enthalpy is a thermodynamic property of a fluid. It can be used to calculate the heat transfer during a quasistatic process taking place in a closed thermodynamic system under constant pressure. Enthalpy H is an arbitrary concept but the Enthalpy change ΔH is more useful because it is equal to the change in the internal energy of the system, plus the work that the system has done on its surroundings.
Sodium	Sodium is a metallic element with a symbol Na and atomic number 11. It is a soft, silvery-white, highly reactive metal and is a member of the alkali metals within "group 1" (formerly known as "group IA".) It has only one stable isotope, ^{23}Na.

Chapter 6. Thermochemistry

Chapter 6. Thermochemistry

Vapor	A Vapor or vapour is a substance in the gas phase at a temperature lower than its critical temperature. This means that the Vapor can be condensed to a liquid or to a solid by increasing its pressure, without reducing the temperature. For example, water has a critical temperature of 374°C (or 647 K) which is the highest temperature at which liquid water can exist.
Vapor pressure	Vapor pressure, is the pressure of a vapor in equilibrium with its non-vapor phases. All liquids and solids have a tendency to evaporate to a gaseous form, and all gases have a tendency to condense back into their original form At any given temperature, for a particular substance, there is a pressure at which the gas of that substance is in dynamic equilibrium with its liquid or solid forms.
Water	Water is the most abundant molecule on Earth"s surface, constituting about 75% of the Earth"s surface. In nature it exists in liquid, solid, and gaseous states. It is in dynamic equilibrium between the liquid and gas states at standard temperature and pressure.
Chemical equation	A Chemical equation may be described as a chemical reaction or a means of writing out and describing such a phenomenon. The coefficients next to the symbols and formulae of entities are the absolute values of the stoichiometric numbers. The first Chemical equation was diagrammed by Jean Beguin in 1615.

Thermochemical equation	A Thermochemical equation is a balanced stoichiometric chemical equation that includes the enthalpy change, ΔH. In variable form, a Thermochemical equation would look like this:
	    A + B → C
	
	    ΔH = (±) #
	Where {A, B, C} are the usual agents of a chemical equation with coefficients and "(±) #" is a positive or negative numerical value, usually with units of kJ.
	Enthalpy (H) is the transfer of energy in a reaction (for chemical reactions it is in the form of heat) and ΔH is the change in enthalpy. ΔH is a state function. Being a state function means that ΔH is independent of the processes between initial and final states.

Ammonia	Ammonia is a compound of nitrogen and hydrogen with the formula NH_3. It is normally encountered as a gas with a characteristic pungent odor. Ammonia contributes significantly to the nutritional needs of terrestrial organisms by serving as a precursor to foodstuffs and fertilizers.
Hydrochloric acid	Hydrochloric acid is the solution of hydrogen chloride (HCl) in water. It is a highly corrosive, strong mineral acid and has major industrial uses. It is found naturally in gastric acid.

Chapter 6. Thermochemistry

Chapter 6. Thermochemistry

Acid	An Acid is traditionally considered any chemical compound that, when dissolved in water, gives a solution with a hydrogen ion activity greater than in pure water, i.e. a pH less than 7.0. That approximates the modern definition of Johannes Nicolaus Brønsted and Martin Lowry, who independently defined an Acid as a compound which donates a hydrogen ion (H^+) to another compound (called a base.) Common examples include acetic Acid and sulfuric Acid (used in car batteries.)
Aqueous solution	An Aqueous solution is a solution in which the solvent is water. It is usually shown in chemical equations by appending (aq) to the relevant formula. The word aqueous means pertaining to, related to, similar to, or dissolved in water.
Hydrazine	Hydrazine is an inorganic chemical compound with the formula N_2H_4. It is a colourless liquid with an ammonia-like odor and is derived from the same industrial chemistry processes that manufacture ammonia. However, Hydrazine has physical properties that are more similar to those of water.
Hydrogen	Hydrogen is the chemical element with atomic number 1. It is represented by the symbol H. At standard temperature and pressure, Hydrogen is a colorless, odorless, nonmetallic, tasteless, highly flammable diatomic gas with the molecular formula H_2. With an atomic weight of 1.007 94 u, Hydrogen is the lightest element.
Solution	In chemistry, a Solution is a homogeneous mixture composed of two or more substances. In such a mixture, a solute is dissolved in another substance, known as a solvent. Gases may dissolve in liquids, for example, carbon dioxide or oxygen in water.
Carbon	Carbon is the chemical element with symbol C and atomic number 6. As a member of group 14 on the periodic table, it is nonmetallic and tetravalent--making four electrons available to form covalent chemical bonds. There are three naturally occurring isotopes, with ^{12}C and ^{13}C being stable, while ^{14}C is radioactive, decaying with a half-life of about 5730 years.
Potassium	Potassium is a chemical element. It has the symbol K , atomic number 19, and atomic mass 39.0983. Potassium was first isolated from potash.
Stoichiometry	Stoichiometry is the calculation of quantitative (measurable) relationships of the reactants and products in a balanced chemical reaction (chemicals.) It can be used to calculate quantities such as the amount of products that can be produced with the given reactants and percent yield. "Stoichiometry" is derived from the Greek words στοιχεά¿–ον and μι̂τρον (metron, meaning measure.)
Sulfur	Sulfur or sulphur is the chemical element that has the atomic number 16. It is denoted with the symbol S. It is an abundant, multivalent non-metal. Sulfur in its native form is a yellow crystalline solid.

Chapter 6. Thermochemistry

Chapter 6. Thermochemistry

White phosphorus	White phosphorus is a flare- and smoke-producing agent and an incendiary agent that is made from a common allotrope of the chemical element phosphorus. The main utility of White phosphorus munitions is to create smokescreens to mask movement from the enemy, or to mask his fire. In contrast to other smoke-causing munitions, White phosphorus burns quickly causing an instant bank of smoke.
Manganese	Manganese is a chemical element, designated by the symbol Mn. It has the atomic number 25. It is found as a free element in nature, and in many minerals.
Phosphorus	Phosphorus is the chemical element that has the symbol P and atomic number 15. A multivalent nonmetal of the nitrogen group, Phosphorus is commonly found in inorganic phosphate rocks. Elemental Phosphorus exists in two major forms - white Phosphorus and red Phosphorus.
Iron	Iron is a chemical element with the symbol Fe and atomic number 26. Iron is a group 8 and period 4 element. Iron and Iron alloys (steels) are by far the most common metals and the most common ferromagnetic materials in everyday use.
Graphite	Graphite " href="/wiki/Cleavage_(crystal)">loose interlamellar coupling between sheets in the structure, in fact in a vacuum environment (such as in technologies for use in space), Graphite was found to be a very poor lubricant. This fact led to the discovery that Graphite"s lubricity is due to adsorbed air and water between the layers, unlike other layered dry lubricants such as molybdenum disulfide. Recent studies suggest that an effect called superlubricity can also account for this effect.
Combustion	Combustion or burning is a complex sequence of exothermic chemical reactions between a fuel (usually a hydrocarbon) and an oxidant accompanied by the production of heat or both heat and light in the form of either a glow or flames, appearance of light flickering. Direct Combustion by atmospheric oxygen is a reaction mediated by radical intermediates. The conditions for radical production are naturally produced by thermal runaway, where the heat generated by Combustion is necessary to maintain the high temperature necessary for radical production.
Tungsten	Tungsten is a chemical element with the chemical symbol W and atomic number 74. A steel-gray metal, Tungsten is found in several ores, including wolframite and scheelite. It is remarkable for its robust physical properties, especially the fact that it has the highest melting point of all the non-alloyed metals and the second highest of all the elements after carbon.
Carbide	In chemistry, a Carbide is a compound composed of carbon and a less electronegative element. Carbide s can be generally classified by chemical bonding type as follows: (i) salt-like, (ii) covalent compounds, (iii) interstitial compounds, and (iv) "intermediate" transition metal Carbide s. Examples include calcium Carbide silicon Carbide tungsten Carbide (often called simply Carbide, and cementite, each used in key industrial applications.

Chapter 6. Thermochemistry

Chapter 6. Thermochemistry

Sublimation	Sublimation of an element or compound is a transition from the solid to gas phase with no intermediate liquid stage. Sublimation is an endothermic phase transition that occurs at temperatures and pressures below the triple point At normal pressures, most chemical compounds and elements possess three different states at different temperatures.
Enthalpy of sublimation	The Enthalpy of sublimation is defined as the heat required to sublime one mole of the substance at a given combination of temperature and pressure, usually standard temperature and pressure (STP.) The heat of sublimation is usually expressed in kJ/mol, although the less customary kJ/kg is also encountered. The standard Enthalpy of sublimation for a material is equivalent to the sum of its standard enthalpy of fusion and its standard enthalpy of vaporization.
Diamond	In mineralogy, Diamond is an allotrope of carbon, where the carbon atoms are arranged in a variation of the face centered cubic crystal structure called a Diamond lattice. Diamond is the second most stable form of carbon, after graphite; however, the conversion rate from Diamond to graphite is negligible at ambient conditions. Diamond is specifically renowned as a material with superlative physical qualities, most of which originate from the strong covalent bonding between its atoms.
Dioxygen	Dioxygen plays an important role in the energy metabolism of living organisms. Free oxygen is produced in the biosphere through photolysis (light-driven oxidation and splitting) of water during photosynthesis in cyanobacteria, green algae, and plants. During oxidative phosphorylation in cellular respiration, oxygen is reduced to water, thus closing the biological water-oxygen redox cycle.
Ozone	Ozone or trioxygen (O_3) is a triatomic molecule, consisting of three oxygen atoms. It is an allotrope of oxygen that is much less stable than the diatomic O_2. Ground-level Ozone is an air pollutant with harmful effects on the respiratory systems of animals.
Activity	In chemical thermodynamics Activity is a measure of the "effective concentration" of a species in a mixture. By convention, it is a dimensionless quantity. The Activity of pure substances in condensed phases (solid or liquids) is normally taken as unity.
Allotropes of Oxygen	There are several known Allotropes of oxygen: - Free Radicals O_1 - unstable - dioxygen, O_2 - colorless - ozone, O_3 - blue - tetraoxygen, O_4 - metastable - solid oxygen exists in 6 variously colored phases - of which one is O_8 and another one metallic

Chapter 6. Thermochemistry

Chapter 6. Thermochemistry

	The common allotrope of elemental oxygen on Earth, O_2, is known as dioxygen. Elemental oxygen is most commonly encountered in this form, as about 21% (by volume) of Earth"s atmosphere. O_2 has a bond length of 121 pm and a bond energy of 498 kJ/mol. Oxygen itself is a colourless gas with a boiling point of -183°C. It can be condensed out of air by cooling with liquid nitrogen, which has a boiling point of -196°C. Liquid oxygen is pale blue in colour, and is quite markedly paramagnetic : liquid oxygen contained in a flask suspended by a string is attracted to a magnet.
Nitric acid	Nitric acid, also known as aqua fortis and spirit of nitre, is a highly corrosive and toxic strong acid that can cause severe burns. Colorless when pure, older samples tend to acquire a yellow cast due to the accumulation of oxides of nitrogen. If the solution contains more than 86% Nitric acid, it is referred to as fuming Nitric acid.
Boiling	Boiling, a type of phase transition, is the rapid vaporization of a liquid, which typically occurs when a liquid is heated to its Boiling point, the temperature at which the vapor pressure of the liquid is equal to the pressure exerted on the liquid by the surrounding environmental pressure. Thus, a liquid may also boil when the pressure of the surrounding atmosphere is sufficiently reduced, such as the use of a vacuum pump or at high altitudes. Boiling occurs in three characteristic stages, which are nucleate, transition and film Boiling.
Freezing	In physical science, Freezing or solidification is the process in which a liquid turns into a solid when cold enough. The Freezing point is the temperature at which this happens. Melting, the process of turning a solid to a liquid, is almost the exact opposite of Freezing.
Gasification	Gasification is a process that converts carbonaceous materials, such as coal, petroleum, biofuel into carbon monoxide and hydrogen by reacting the raw material, such as house waste, or compost at high temperatures with a controlled amount of oxygen and/or steam. The resulting gas mixture is called synthesis gas or syngas and is itself a fuel. Gasification is a method for extracting energy from many different types of organic materials.
Coal Liquefaction	Coal liquefaction is the process of producing synthetic liquid fuels from coal. The liquefaction processes are classified as direct conversion to liquids processes and indirect conversion to liquids processeses. Direct processes are carbonization and hydrogenation.
Acid rain	Acid rain is rain or any other form of precipitation that is unusually acidic. It has harmful effects on plants, aquatic animals, and infrastructure. Acid rain is mostly caused by human emissions of sulfur and nitrogen compounds which react in the atmosphere to produce acids.
Liquid oxygen	Liquid oxygen is a form of the element oxygen. It has a pale blue color and is strongly paramagnetic. Liquid oxygen has a density of 1.141 g/cmÂ³ (1.141 kg/L) and is moderately cryogenic (freezing point: 50.5 K (−222.65 °C), boiling point: 90.188 K (−182.96 °C) at 101.325 kPa (760 mm Hg.)

Chapter 6. Thermochemistry

Chapter 6. Thermochemistry

Metallurgy | Metallurgy is a domain of materials science that studies the physical and chemical behavior of metallic elements, their intermetallic compounds, and their mixtures, which are called alloys. It is also the technology of metals: the way in which science is applied to their practical use. Metallurgy is commonly used in the craft of metalworking.

Chapter 6. Thermochemistry

Chapter 7. Quantum Theory of the Atom

Copper	Copper is a chemical element with the symbol Cu and atomic number 29. It is a ductile metal with very high thermal and electrical conductivity. Pure Copper is rather soft and malleable and a freshly-exposed surface has a pinkish or peachy color.
Emission	In physics, Emission is the process by which the energy of a photon is released by another entity, for example, by an atom whose electrons make a transition between two electronic energy levels. The emitted energy is in the form of a photon. The emittance of an object quantifies how much light is emitted by it.
Linear	In chemistry, the Linear molecular geometry describes the arrangement of three or more atoms placed at an expected bond angle of 180°. Linear organic molecules, e.g. acetylene, are often described by invoking sp orbital hybridization for the carbon centers. Many Linear molecules exist, prominent examples include CO_2, HCN, and xenon difluoride.
Lithium	Lithium is the chemical element with atomic number 3, and is represented by the symbol Li. It is a soft alkali metal with a silver-white color. Under standard conditions it is the lightest metal and the least dense solid element.
Metal	In chemistry, a Metal is an element, compound, or alloy characterized by high electrical conductivity. In a Metal atoms readily lose electrons to form positive ions ; those ions are surrounded by delocalized electrons, which are responsible for the conductivity. The thus produced solid is held by electrostatic interactions between the ions and the electron cloud, which are called Metal lic bonds.
Rutherford	The Rutherford is an obsolete unit of radioactivity, defined as the activity of a quantity of radioactive material in which one million nuclei decay per second. It is therefore equivalent to one megabecquerel. It was named after Ernest Rutherford It is not an SI unit.
Electron	The Electron is a subatomic particle that carries a negative electric charge. It has no known substructure and is believed to be a point particle. An Electron has a mass that is approximately 1836 times less than that of the proton.
Electron pair	In chemistry, an Electron pair consists of two electrons that occupy the same orbital but have opposite spins. MO diagrams depicting covalent (left) and polar covalent (right) bonding in a diatomic molecule. In both cases a bond is created by the formation of an Electron pair. Because electrons are fermions, the Pauli exclusion principle forbids these particles from having exactly the same quantum numbers.

Chapter 7. Quantum Theory of the Atom

Chapter 7. Quantum Theory of the Atom

Electrolysis

In chemistry and manufacturing, Electrolysis is a method of using an electric current to drive an otherwise non-spontaneous chemical reaction. Electrolysis is commercially highly important as a stage in the separation of elements from naturally-occurring sources such as ores using an electrolytic cell.

- 1800 - William Nicholson and Johann Ritter decomposed water into hydrogen and oxygen.
- 1807 - Potassium, Sodium, Barium, Calcium and Magnesium were discovered by Sir Humphry Davy using Electrolysis.
- 1886 - Fluorine was discovered by Henri Moissan using Electrolysis.
- 1886 - Hall-Héroult process developed for making aluminium
- 1890 - Castner-Kellner process developed for making sodium hydroxide

Electrolysis is the passage of an electric current through an ionic substance that is either molten or dissolved in a suitable solvent, resulting in chemical reactions at the electrodes and separation of materials.

The main components required to achieve Electrolysis are:

- A liquid containing mobile ions - an electrolyte
- An external source of direct electric current
- Two solid rods or plates known as electrodes

The components perform the following roles in the Electrolysis process:

- The mobile ions are the carriers of electrical current in the liquid (electrolyte.) If the ions are not mobile, as in a solid salt then Electrolysis cannot occur.

Diffraction

Diffraction is normally taken to refer to various phenomena which occur when a wave encounters an obstacle. It is described as the apparent bending of waves around small obstacles and the spreading out of waves past small openings. Very similar effects are observed when there is an alteration in the properties of the medium in which the wave is travelling, for example a variation in refractive index for light waves or in acoustic impedance for sound waves and these can also be referred to as Diffraction effects.

Einstein

An Einstein is a unit used in irradiance and in photochemistry. One Einstein is defined as one mole of photons, regardless of their frequency. Therefore, the number of photons in an Einstein is Avogadro"s number.

Chapter 7. Quantum Theory of the Atom

Chapter 7. Quantum Theory of the Atom

Photoelectric effect	The Photoelectric effect is a phenomenon in which electrons are emitted from matter (metals and non-metallic solids, liquids, or gases) after the absorption of energy from electromagnetic radiation such as X-rays or visible light. The emitted electrons can be referred to as photoelectrons in this context. The effect is also termed the Hertz Effect, due to its discovery by Heinrich Rudolf Hertz, although the term has generally fallen out of use.
Wave-particle duality	In physics and chemistry, wave-particle duality is the concept that all matter and energy exhibits both wave-like and particle-like properties. A central concept of quantum mechanics, duality addresses the inadequacy of classical concepts like "particle" and "wave" in fully describing the behaviour of small-scale objects. Various interpretations of quantum mechanics attempt to explain this ostensible paradox.
Vapor	A Vapor or vapour is a substance in the gas phase at a temperature lower than its critical temperature. This means that the Vapor can be condensed to a liquid or to a solid by increasing its pressure, without reducing the temperature. For example, water has a critical temperature of 374°C (or 647 K) which is the highest temperature at which liquid water can exist.
Vapor pressure	Vapor pressure, is the pressure of a vapor in equilibrium with its non-vapor phases. All liquids and solids have a tendency to evaporate to a gaseous form, and all gases have a tendency to condense back into their original form At any given temperature, for a particular substance, there is a pressure at which the gas of that substance is in dynamic equilibrium with its liquid or solid forms.
Atom	The Atom is a basic unit of matter consisting of a dense, central nucleus surrounded by a cloud of negatively charged electrons. The atomic nucleus contains a mix of positively charged protons and electrically neutral neutrons (except in the case of hydrogen-1, which is the only stable nuclide with no neutron.) The electrons of an Atom are bound to the nucleus by the electromagnetic force.
Hydrogen	Hydrogen is the chemical element with atomic number 1. It is represented by the symbol H. At standard temperature and pressure, Hydrogen is a colorless, odorless, nonmetallic, tasteless, highly flammable diatomic gas with the molecular formula H_2. With an atomic weight of 1.007 94 u, Hydrogen is the lightest element.
Hydrogen atom	A Hydrogen atom is an atom of the chemical element hydrogen, and an example of a Boson. The electrically neutral atom contains a single positively-charged proton and a single negatively-charged electron bound to the nucleus by the Coulomb force. The most abundant isotope, hydrogen-1, protium, or light hydrogen, contains no neutrons; other isotopes contain one or more neutrons.

Chapter 7. Quantum Theory of the Atom

Chapter 7. Quantum Theory of the Atom

Cobalt	Cobalt is a hard, lustrous, grey metal, a chemical element with symbol Co and atomic number 27. Although Cobalt based colors and pigments have been used since ancient times for making jewelry and paints, and miners have long used the name kobold ore for some minerals, the free metallic Cobalt was not prepared and discovered until 1735 by Georg Brandt.
Catalyst	Catalyst is a science centre and museum devoted to the chemical industry. Its full title is Catalyst Science Discovery Centre. It is located in Widnes, Cheshire, in the north west of England, and situated on the north bank of the River Mersey (grid reference SJ512841.)
Continuous spectrum	The position operator usually has a Continuous spectrum, much like the momentum operator in an infinite space. But the momentum in a compact space, the angular momentum, and the Hamiltonian of various physical systems, specially bound states, tend to have a discrete (quantized) spectrum -- that is where the name quantum mechanics comes from. However computing the spectra or cross sections associated with scattering experiments (like for instance high resolution electron energy loss spectroscopy) usually requires the computation of the non quantized or Continuous spectrum of the Hamiltonian.
Hydrogen line	The Hydrogen line, 21 centimeter line or HI line refers to the spectral line created by changing in the energy state of neutral hydrogen and occurs at a frequency of 1420.40575177 MHz, equivalent to a vacuum wavelength of around 21.10611405413 cm. This line falls within the microwave region of the electromagnetic spectrum and is used extensively in astronomy, since it can penetrate dust clouds that are opaque to visible wavelengths. The radiation comprising the Hydrogen line comes from the transition between the two hyperfine levels of the hydrogen 1s ground state.
Energy levels	A quantum mechanical system or particle that is bound, confined spatially, can only take on certain discrete values of energy, as opposed to classical particles, which can have any energy. These values are called Energy levels. The term is most commonly used for the Energy levels of electrons in atoms or molecules, which are bound by the electric field of the nucleus.
Pauli exclusion principle	The Pauli exclusion principle is a quantum mechanical principle formulated by Wolfgang Pauli in 1925. It states that no two identical fermions may occupy the same quantum state simultaneously. A more rigorous statement of this principle is that, for two identical fermions, the total wave function is anti-symmetric.
Principal quantum number	In atomic physics, the Principal quantum number symbolized as n is the first of a set of quantum numbers (which includes: the Principal quantum number, the azimuthal quantum number, the magnetic quantum number, and the spin quantum number) of an atomic orbital.
	There are a set of quantum numbers associated with the energy states of the atom. The four quantum numbers n, â„", m, and s specify the complete and unique quantum state of a single electron in an atom called its wavefunction or orbital.

Chapter 7. Quantum Theory of the Atom

Chapter 7. Quantum Theory of the Atom

Atomic orbital	An Atomic orbital is a mathematical function that describes the wave-like behavior of either one electron or a pair of electrons, in an atom. This function can be used to calculate the probability of finding any electron of an atom in any specific region around the atom"s nucleus. These functions may serve as three-dimensional graph of an electron"s likely location.
Principle	A Principle is a comprehensive and fundamental law, doctrine, or assumption. It can be a rule or code of conduct. It can be a law or fact of nature underlying the working of an artificial device.
Chromium	Chromium is a chemical element which has the symbol Cr and atomic number 24. It is a steely-gray, lustrous, hard metal that takes a high polish and has a high melting point. It is also odourless, tasteless, and malleable.
Ion	An Ion is an atom or molecule where the total number of electrons is not equal to the total number of protons, giving it a net positive or negative electrical charge. Since protons are positively charged and electrons are negatively charged, if there are more electrons than protons, the atom or molecule will be negatively charged. This is called an an Ion , from the Greek á¼€vÎ¬ , meaning "up".
Electron microscope	An Electron microscope is a type of microscope that uses a particle beam of electrons to illuminate a specimen and create a highly-magnified image. Electron microscope s have much greater resolving power than light microscopes that use electromagnetic radiation and can obtain much higher magnifications of up to 2 million times, while the best light microscopes are limited to magnifications of 2000 times. Both electron and light microscopes have resolution limitations, imposed by the wavelength of the radiation they use.
Microscopy	Microscopy is the technical field of using microscopes to view samples or objects. There are three well-known branches of Microscopy, optical, electron and scanning probe Microscopy. Optical and electron Microscopy involve the diffraction, reflection, or refraction of electromagnetic radiation/electron beam interacting with the subject of study, and the subsequent collection of this scattered radiation in order to build up an image.
Scanning electron microscope	The Scanning electron microscope is a type of electron microscope that images the sample surface by scanning it with a high-energy beam of electrons in a raster scan pattern. The electrons interact with the atoms that make up the sample producing signals that contain information about the sample"s surface topography, composition and other properties such as electrical conductivity. The types of signals produced by an Scanning electron microscope include secondary electrons, back scattered electrons (BSE), characteristic x-rays, light (cathodoluminescence), specimen current and transmitted electrons.

Chapter 7. Quantum Theory of the Atom

Chapter 7. Quantum Theory of the Atom

Microscope	A Microscope is an instrument for viewing objects that are too small to be seen by the naked or unaided eye. The science of investigating small objects using such an instrument is called microscopy. The term microscopic means minute or very small, not visible with the eye unless aided by a Microscope.
Azimuthal quantum number	The Azimuthal quantum number symbolized as â„" (lower-case L) is a quantum number for an atomic orbital that determines its orbital angular momentum. The Azimuthal quantum number is the second of a set of quantum numbers (the principal quantum number, following spectroscopic notation, the Azimuthal quantum number, the magnetic quantum number, and the spin quantum number) which describe the unique quantum state of an electron and is designated by the letter â„".
	There is a set of quantum numbers associated with the energy states of the electrons of an atom.
Magnetic quantum number	In atomic physics, the Magnetic quantum number is the third of a set of quantum numbers (the principal quantum number, the azimuthal quantum number, the Magnetic quantum number, and the spin quantum number) which describe the unique quantum state of an electron and is designated by the letter m. The Magnetic quantum number denotes the energy levels available within a subshell.
	There are a set of quantum numbers associated with the energy states of the atom.
Spin quantum number	In atomic physics, the Spin quantum number is a quantum number that parameterizes the intrinsic angular momentum (or spin angular momentum, or simply spin) of a given particle. The Spin quantum number is the fourth of a set of quantum numbers (the principal quantum number, the azimuthal quantum number, the magnetic quantum number, and the Spin quantum number) which describe the unique quantum state of an electron and is designated by the letter s.
	As a quantized angular momentum, it holds that $$\|\mathbf{s}\| = \sqrt{s(s+1)}\,\hbar$$ where $\mathbf{S}$ is the quantized spin vector, $\|\mathbf{s}\|$ is the norm of the spin vector, s is the Spin quantum number associated with the spin angular momentum, $\hbar$ is the reduced Planck constant.

Chapter 7. Quantum Theory of the Atom

Chapter 8. Electron Configurations and Periodicity

Beryllium	Beryllium is the chemical element with the symbol Be and atomic number 4.
	A bivalent element, Beryllium is found naturally only combined with other elements in minerals. Notable gemstones which contain Beryllium include Beryl (aquamarines and emeralds) and Chrysoberyl (Alexandrite and Cat"s eye.)
Calcium	Calcium is the chemical element with the symbol Ca and atomic number 20. It has an atomic mass of 40.078 amu. Calcium is a soft grey alkaline earth metal, and is the fifth most abundant element by mass in the Earth"s crust.
Curie	The Curie is a unit of radioactivity, defined as
	$\quad$ 1 Ci = 3.7×10^{10} decays per second or becquerels.
	This is roughly the activity of 1 gram of the radium isotope ^{226}Ra, a substance studied by the pioneers of radiology, Marie and Pierre Curie. The Curie has since been replaced by an SI derived unit, the becquerel (Bq), which equates to one decay per second.
Marie SkÅ‚odowska Curie	Marie SkÅ‚odowska Curie was a physicist and chemist of Polish upbringing and, subsequently, French citizenship. She was a pioneer in the field of radioactivity, the first person honored with two Nobel Prizes, and the first female professor at the University of Paris.
	She was born Maria SkÅ‚odowska in Warsaw and lived there until she was 24.
Pierre Curie	Pierre Curie was a French physicist, a pioneer in crystallography, magnetism, piezoelectricity and radioactivity, and Nobel laureate. In 1903 he received the Nobel Prize in Physics with his wife, Maria SkÅ‚odowska-Curie, and Henri Becquerel, "in recognition of the extraordinary services they have rendered by their joint researches on the radiation phenomena discovered by Professor Henri Becquerel."
	Born in Paris, France, Pierre was educated by his father, and in his early teens showed a strong aptitude for mathematics and geometry. By the age of 18 he had completed the equivalent of a higher degree, but did not proceed immediately to a doctorate due to lack of money.
Electron	The Electron is a subatomic particle that carries a negative electric charge. It has no known substructure and is believed to be a point particle. An Electron has a mass that is approximately 1836 times less than that of the proton.
Electron configuration	In atomic physics and quantum chemistry, Electron configuration is the arrangement of electrons of an atom, a molecule, or other physical structure. It concerns the way electrons can be distributed in the orbitals of the given system (atomic or molecular for instance.)
	Like other elementary particles, the electron is subject to the laws of quantum mechanics, and exhibits both particle-like and wave-like nature.

Chapter 8. Electron Configurations and Periodicity

Chapter 8. Electron Configurations and Periodicity

Periodic table

The periodiÑ table of the chemical elements (also, Periodic table of the elements or just Periodic table) is a tabular display of the chemical elements. Although precursors to this table exist, its invention is generally credited to Russian chemist Dmitri Mendeleev in 1869, who intended the table to illustrate recurring trends in the properties of the elements. The layout of the table has been refined and extended over time, as new elements have been discovered, and new theoretical models have been developed to explain chemical behavior.

Radium

Radium is a radioactive chemical element which has the symbol Ra and atomic number 88. Its appearance is almost pure white, but it readily oxidizes on exposure to air, turning black. Radium is an alkaline earth metal that is found in trace amounts in uranium ores.

Spin quantum number

In atomic physics, the Spin quantum number is a quantum number that parameterizes the intrinsic angular momentum (or spin angular momentum, or simply spin) of a given particle. The Spin quantum number is the fourth of a set of quantum numbers (the principal quantum number, the azimuthal quantum number, the magnetic quantum number, and the Spin quantum number) which describe the unique quantum state of an electron and is designated by the letter s.

As a quantized angular momentum, it holds that

$$\|\mathbf{s}\| = \sqrt{s(s+1)}\,\hbar$$

where

$\mathbf{S}$ is the quantized spin vector,
$\|\mathbf{s}\|$ is the norm of the spin vector,
s is the Spin quantum number associated with the spin angular momentum,
$\hbar$ is the reduced Planck constant.

Configuration

The configuration of a molecule is the permanent geometry that results from the spatial arrangement of its bonds. The ability of the same set of atoms to form two or more molecules with different configurations is stereoisomerism. configuration is distinct from chemical conformation, a shape attainable by bond rotations.

Lithium

Lithium is the chemical element with atomic number 3, and is represented by the symbol Li. It is a soft alkali metal with a silver-white color. Under standard conditions it is the lightest metal and the least dense solid element.

Pauli exclusion principle

The Pauli exclusion principle is a quantum mechanical principle formulated by Wolfgang Pauli in 1925. It states that no two identical fermions may occupy the same quantum state simultaneously. A more rigorous statement of this principle is that, for two identical fermions, the total wave function is anti-symmetric.

Chapter 8. Electron Configurations and Periodicity

Chapter 8. Electron Configurations and Periodicity

Principle	A Principle is a comprehensive and fundamental law, doctrine, or assumption. It can be a rule or code of conduct. It can be a law or fact of nature underlying the working of an artificial device.
Nuclear magnetic resonance	Nuclear magnetic resonance is a property that magnetic nuclei have in a magnetic field and applied electromagnetic (EM) pulse, which cause the nuclei to absorb energy from the EM pulse and radiate this energy back out. The energy radiated back out is at a specific resonance frequency which depends on the strength of the magnetic field and other factors. This allows the observation of specific quantum mechanical magnetic properties of an atomic nucleus.
Resonance	Resonance in chemistry is a key component of valence bond theory used to graphically represent and mathematically model certain types of molecular structures when no single, conventional Lewis structure can satisfactorily represent the observed structure or explain its properties. Resonance instead considers such molecules to be an intermediate or average (called a Resonance hybrid) between several Lewis structures that differ only in the placement of the valence electrons. Scheme 1.
Chemical shift	In nuclear magnetic resonance (NMR), the Chemical shift describes the dependence of nuclear magnetic energy levels on the electronic environment in a molecule. Chemical shift s are relevant in NMR spectroscopy techniques such as proton NMR and carbon-13 NMR. An atomic nucleus can have a magnetic moment (nuclear spin), which gives rise to different energy levels and resonance frequencies in a magnetic field. The total magnetic field experienced by a nucleus includes local magnetic fields induced by currents of electrons in the molecular orbitals (note that electrons have a magnetic moment themselves.)
Sodium	Sodium is a metallic element with a symbol Na and atomic number 11. It is a soft, silvery-white, highly reactive metal and is a member of the alkali metals within "group 1" (formerly known as "group IA".) It has only one stable isotope, ^{23}Na.
Aufbau principle	The Aufbau principle is used to determine the electron configuration of an atom, molecule or ion. The principle postulates a hypothetical process in which an atom is "built up" by progressively adding electrons. As they are added, they assume their most stable conditions with respect to the nucleus and those electrons already there.
Argon	Argon is a chemical element designated by the symbol Ar. Argon has atomic number 18 and is the third element in group 18 of the periodic table . Argon is present in the Earth"s atmosphere at 0.94%.
Boron	Boron is the chemical element with atomic number 5 and the chemical symbol B. Boron is a trivalent metalloid element which occurs abundantly in the evaporite ores borax and ulexite. Several allotropes of Boron exist; amorphous Boron is a brown powder, though crystalline Boron is black, extremely hard , and a poor conductor at room temperature. Elemental Boron is used as a dopant in the semiconductor industry, while Boron compounds play important roles as light structural materials, insecticides and preservatives, and reagents for chemical synthesis.

Chapter 8. Electron Configurations and Periodicity

Chapter 8. Electron Configurations and Periodicity

Carbon	Carbon is the chemical element with symbol C and atomic number 6. As a member of group 14 on the periodic table, it is nonmetallic and tetravalent--making four electrons available to form covalent chemical bonds. There are three naturally occurring isotopes, with ^{12}C and ^{13}C being stable, while ^{14}C is radioactive, decaying with a half-life of about 5730 years.
Gallium	Gallium is a chemical element that has the symbol Ga and atomic number 31. Elemental Gallium does not occur in nature, but as the Ga salt, in trace amounts in bauxite and zinc ores. A soft silvery metallic poor metal, elemental Gallium is a brittle solid at low temperatures.
Helium	Helium is the chemical element with atomic number 2, and is represented by the symbol He. It is a colorless, odorless, tasteless, non-toxic, inert monatomic gas that heads the noble gas group in the periodic table. Its boiling and melting points are the lowest among the elements and it exists only as a gas except in extreme conditions.
Hydrogen	Hydrogen is the chemical element with atomic number 1. It is represented by the symbol H. At standard temperature and pressure, Hydrogen is a colorless, odorless, nonmetallic, tasteless, highly flammable diatomic gas with the molecular formula H_2. With an atomic weight of 1.007 94 u, Hydrogen is the lightest element.
Magnesium	Magnesium is a chemical element with the symbol Mg, atomic number 12, atomic weight 24.3050 and common oxidation number +2. Magnesium, an alkaline earth metal, is the ninth most abundant element in the universe by mass. The commonness of Magnesium is related to the fact that it is easily built up in supernova stars from a sequential addition of three helium nuclei to carbon .
Neon	Neon is the chemical element that has the symbol Ne and atomic number 10. Although a very common element in the universe, it is rare on Earth. A colorless, inert noble gas under standard conditions, Neon gives a distinct reddish-orange glow when used in discharge tubes and Neon lamps.
Potassium	Potassium is a chemical element. It has the symbol K , atomic number 19, and atomic mass 39.0983. Potassium was first isolated from potash.
Scandium	Scandium is a chemical element with symbol Sc and atomic number 21. A silvery-white metallic transition metal, it has historically been sometimes classified as a rare earth element, together with yttrium and the lanthanides. In 1879 Lars Fredrik Nilson and his team, found a new element with spectral analysis, in the minerals euxenite and gadolinite from Scandinavia.

Chapter 8. Electron Configurations and Periodicity

Chapter 8. Electron Configurations and Periodicity

Titanium	Titanium is a chemical element with the symbol Ti and atomic number 22. Sometimes called the "space age metal", it has a low density and is a strong, lustrous, corrosion-resistant transition metal with a silver color. Titanium can be alloyed with iron, aluminium, vanadium, molybdenum, among other elements, to produce strong lightweight alloys for aerospace (jet engines, missiles, and spacecraft), military, industrial process (chemicals and petro-chemicals, desalination plants, pulp, and paper), automotive, agri-food, medical prostheses, orthopedic implants, dental and endodontic instruments and files, dental implants, sporting goods, jewelry, mobile phones, and other applications.
Vanadium	Vanadium is the chemical element with the symbol V and atomic number 23. It is a soft, silvery grey, ductile transition metal. The formation of an oxide layer stabilizes the metal against oxidation.
Zinc	Zinc is a metallic chemical element with the symbol Zn and atomic number 30. It is a first-row transition metal in group 12 of the periodic table. Zinc is chemically similar to magnesium because its ion is of similar size and its only common oxidation state is +2.
Abundance	The abundance of a chemical element measures how relatively common the element is, or how much of the element there is by comparison to all other elements. abundance may be variously measured by the mass-fraction (the same as weight fraction), or mole-fraction (fraction of atoms, or sometimes fraction of molecules, in gases), or by volume fraction. Measurement by volume-fraction is a common abundance measure in mixed gases such as atmospheres, which is close to molecular mole-fraction for ideal gas mixtures (i.e., gas mixtures at relatively low densities and pressures.)
Acid	An Acid is traditionally considered any chemical compound that, when dissolved in water, gives a solution with a hydrogen ion activity greater than in pure water, i.e. a pH less than 7.0. That approximates the modern definition of Johannes Nicolaus Brønsted and Martin Lowry, who independently defined an Acid as a compound which donates a hydrogen ion (H^+) to another compound (called a base.) Common examples include acetic Acid and sulfuric Acid (used in car batteries.)
Atomic mass	The Atomic mass is the mass of an atom, most often expressed in unified Atomic mass units. The Atomic mass may be considered to be the total mass of protons, neutrons and electrons in a single atom (when the atom is motionless.) The Atomic mass is sometimes incorrectly used as a synonym of relative Atomic mass, average Atomic mass and atomic weight; however, these differ subtly from the Atomic mass.
Covalent radius	The Covalent radius, r_{cov}, is a measure of the size of atom that forms part of a covalent bond. It is measured either in picometres (pm) or ångströms (Å), with 1 Å = 100 pm. In principle, the sum of the two covalent radii should equal the covalent bond length between two atoms.

Chapter 8. Electron Configurations and Periodicity

Chapter 8. Electron Configurations and Periodicity

Crystal	A Crystal or Crystal line solid is a solid material whose constituent atoms, molecules, or ions are arranged in an orderly repeating pattern extending in all three spatial dimensions. The scientific study of Crystal s and Crystal formation is Crystal lography. The process of Crystal formation via mechanisms of Crystal growth is called Crystal lization or solidification.
Crystal structure	In mineralogy and crystallography, a Crystal structure is a unique arrangement of atoms in a crystal. A Crystal structure is composed of a motif, a set of atoms arranged in a particular way, and a lattice. Motifs are located upon the points of a lattice, which is an array of points repeating periodically in three dimensions.
Alkaline earth metals	The Alkaline earth metals are a series of elements comprising Group 2 (IUPAC style) (Group IIA) of the periodic table: beryllium (Be), magnesium (Mg), calcium (Ca), strontium (Sr), barium (Ba) and radium (Ra.) This specific group in the periodic table owes its name to their oxides that simply give basic alkaline solutions. These elements melt at such high temperature that they remain solids ("earths") in fires.
Group	In chemistry, a Group is a vertical column in the periodic table of the chemical elements. The name family is derived from the fact that the elements share similar characteristics and traits, just as members of any human family would. There are 18 groups in the standard periodic table.
Krypton	Krypton is a chemical element with the symbol Kr and atomic number 36. It is a member of Group 18 and Period 4 elements. A colorless, odorless, tasteless noble gas, Krypton occurs in trace amounts in the atmosphere, is isolated by fractionally distilling liquified air, and is often used with other rare gases in fluorescent lamps.
Noble gas	The Noble gas es are a group of chemical elements with very similar properties: under standard conditions, they are all odorless, colorless, monatomic gases, with a very low chemical reactivity. The six Noble gas es that occur naturally are helium (He), neon (Ne), argon (Ar), krypton (Kr), xenon (Xe), and the radioactive radon (Rn.) For the first six periods of the periodic table, the Noble gas es are exactly the members of group 18 of the periodic table.
Valence	In chemistry, Valence is a measure of the number of chemical bonds formed by the atoms of a given element. Over the last century, the concept of Valence evolved into a range of approaches for describing the chemical bond, including Lewis structures, Valence bond theory, molecular orbitals, Valence shell electron pair repulsion theory and all the advanced methods of quantum chemistry. The etymology of the word "Valence" is from 1425, meaning "extract, preparation," from Latin valentia "strength, capacity," and the chemical meaning referring to the "combining power of an element" is recorded from 1884, from German Valenz.

Chapter 8. Electron Configurations and Periodicity

Chapter 8. Electron Configurations and Periodicity

Valence electrons	In science, Valence electrons are the outermost electrons of an atom, which are important in determining how the atom reacts chemically with other atoms. Atoms with a complete shell of Valence electrons tend to be chemically inert. Atoms with one or two Valence electrons more than a closed shell are highly reactive because the extra electrons are easily removed to form positive ions.
Atomic orbital	An Atomic orbital is a mathematical function that describes the wave-like behavior of either one electron or a pair of electrons, in an atom. This function can be used to calculate the probability of finding any electron of an atom in any specific region around the atom"s nucleus. These functions may serve as three-dimensional graph of an electron"s likely location.
D-block	The D-block of the periodic table of the elements consists of those periodic table groups that contain elements in which, in the atomic ground state, the highest-energy electron is in a d-orbital. The D-block elements are often also known as transition metals. Although Lutetium and Lawrencium are in the D-block, they are not considered transition metals but a lanthanide and an actinide, respectively, according to IUPAC. Group 12 elements are also in the D-block but are considered post-transition metals if their d-subshell is completely filled.
Earth	Earth is the third planet from the Sun, and the largest of the terrestrial planets in the Solar System in terms of diameter, mass and density. It is also referred to as the World, the Blue Planet, and Terra. Home to millions of species, including humans, Earth is the only place in the universe where life is known to exist.
F-block	The F-block of the periodic table of the elements consists of those elements (sometimes referred to as the inner transition elements or rare earth metals) for which, in the atomic ground state, the highest-energy electrons occupy f-orbitals. Unlike the other blocks, the conventional divisions of the F-block follow periods of similar atomic number rather than groups of similar electron configuration. Thus, the F-block is divided into the lanthanoid series and the actinoid series.
Isomer	In chemistry, Isomer s are compounds with the same molecular formula but different structural formula. Isomer s do not necessarily share similar properties unless they also have the same functional groups. This should not be confused with a nuclear Isomer which involves a nucleus at different states of excitement.
Metal	In chemistry, a Metal is an element, compound, or alloy characterized by high electrical conductivity. In a Metal atoms readily lose electrons to form positive ions ; those ions are surrounded by delocalized electrons, which are responsible for the conductivity. The thus produced solid is held by electrostatic interactions between the ions and the electron cloud, which are called Metal lic bonds.

Chapter 8. Electron Configurations and Periodicity

Chapter 8. Electron Configurations and Periodicity

Chromium	Chromium is a chemical element which has the symbol Cr and atomic number 24. It is a steely-gray, lustrous, hard metal that takes a high polish and has a high melting point. It is also odourless, tasteless, and malleable.
Copper	Copper is a chemical element with the symbol Cu and atomic number 29. It is a ductile metal with very high thermal and electrical conductivity. Pure Copper is rather soft and malleable and a freshly-exposed surface has a pinkish or peachy color.
Photoelectric effect	The Photoelectric effect is a phenomenon in which electrons are emitted from matter (metals and non-metallic solids, liquids, or gases) after the absorption of energy from electromagnetic radiation such as X-rays or visible light. The emitted electrons can be referred to as photoelectrons in this context. The effect is also termed the Hertz Effect, due to its discovery by Heinrich Rudolf Hertz, although the term has generally fallen out of use.
X-ray photoelectron spectroscopy	X-ray photoelectron spectroscopy is a quantitative spectroscopic technique that measures the elemental composition, empirical formula, chemical state and electronic state of the elements that exist within a material. X-ray photoelectron spectroscopy spectra are obtained by irradiating a material with a beam of X-rays while simultaneously measuring the kinetic energy (KE) and number of electrons that escape from the top 1 to 10 nm of the material being analyzed. X-ray photoelectron spectroscopy requires ultra high vacuum (UHV) conditions.
Spectroscopy	Spectroscopy was originally the study of the interaction between radiation and matter as a function of wavelength (λ.) In fact, historically, Spectroscopy referred to the use of visible light dispersed according to its wavelength, e.g. by a prism. Later the concept was expanded greatly to comprise any measurement of a quantity as function of either wavelength or frequency.
Ionization	Ionization is the physical process of converting an atom or molecule into an ion by adding or removing charged particles such as electrons or other ions. This is often confused with dissociation (chemistry.) The process works slightly differently depending on whether an ion with a positive or a negative electric charge is being produced.
Bromine	Bromine , Greek: βρά¿¶μος, brómos, meaning "stench "), is a chemical element with the symbol Br and atomic number 35. A halogen element, Bromine is a reddish-brown volatile liquid at standard room temperature that is intermediate in reactivity between chlorine and iodine. Bromine vapours are corrosive and toxic.
Tellurium	Tellurium is a chemical element that has the symbol Te and atomic number 52. A brittle silver-white metalloid which looks like tin, Tellurium is chemically related to selenium and sulfur. Tellurium is primarily used in alloys and as a semiconductor.

Chapter 8. Electron Configurations and Periodicity

Chapter 8. Electron Configurations and Periodicity

Oxygen	Oxygen and -γενῖ®ς (-genÄ"s) (producer, literally begetter) is the element with atomic number 8 and represented by the symbol O. It is a member of the chalcogen group on the periodic table, and is a highly reactive nonmetallic period 2 element that readily forms compounds (notably oxides) with almost all other elements. At standard temperature and pressure two atoms of the element bind to form di Oxygen , a colorless, odorless, tasteless diatomic gas with the formula O_2. Oxygen is the third most abundant element in the universe by mass after hydrogen and helium and the most abundant element by mass in the Earth"s crust.
Atomic weight	Atomic weight (symbol: A_r) is a dimensionless physical quantity, the ratio of the average mass of atoms of an element to 1/12 of the mass of an atom of carbon-12. The term is usually used, without further qualification, to refer to the standard Atomic weight s published at regular intervals by the International Union of Pure and Applied Chemistry (IUPAC) and which are intended to be applicable to normal laboratory materials
Iron	Iron is a chemical element with the symbol Fe and atomic number 26. Iron is a group 8 and period 4 element. Iron and Iron alloys (steels) are by far the most common metals and the most common ferromagnetic materials in everyday use.
Atom	The Atom is a basic unit of matter consisting of a dense, central nucleus surrounded by a cloud of negatively charged electrons. The atomic nucleus contains a mix of positively charged protons and electrically neutral neutrons (except in the case of hydrogen-1, which is the only stable nuclide with no neutron.) The electrons of an Atom are bound to the nucleus by the electromagnetic force.
Periodic trends	In Chemistry, Periodic trends are the tendencies of certain elemental characteristics to increase or decrease as one progresses from one corner of the Periodic table of elements. The atomic radius is the distance from the atomic nucleus to the outermost stable electron orbital in an atom that is at equilibrium. The atomic radius tends to decrease as one progresses across a period because the effective nuclear charge increases, thereby attracting the orbiting electrons and lessening the radius.
Activity	In chemical thermodynamics Activity is a measure of the "effective concentration" of a species in a mixture. By convention, it is a dimensionless quantity. The Activity of pure substances in condensed phases (solid or liquids) is normally taken as unity.
Oxidation number	The Oxidation number of a central atom in a coordination compound is the charge that it would have if all the ligands were removed along with the electron pairs that were shared with the central atom. It is used in the nomenclature of inorganic compounds. It is represented by a Roman numeral; the plus sign is omitted for positive Oxidation number s.

Chapter 8. Electron Configurations and Periodicity

Chapter 8. Electron Configurations and Periodicity

Germanium	Germanium is a chemical element with the symbol Ge and atomic number 32. It is a lustrous, hard, grayish-white metalloid in the carbon group, chemically similar to its group neighbors tin and silicon. Germanium has five naturally occurring isotopes ranging in atomic mass number from 70 to 76.
Molecular orbital	In chemistry, a Molecular orbital is a mathematical function that describes the wave-like behavior of an electron in a molecule. This function can be used to calculate chemical and physical properties such as the probability of finding an electron in any specific region. The use of the term "orbital" was first used in English by Robert S. Mulliken in 1925 as the English translation of Schrödinger"s use of the German word, "Eigenfunktion".
Atomic radius	Atomic radius, is called the width of an atom, but it is not a precisely defined physical quantity, nor is it constant in all circumstances. The value assigned to the radius of a particular atom always depends on the definition chosen for "Atomic radius," and the appropriate definition depends on the context. The term "Atomic radius" itself is problematic: it may be restricted to the size of free atoms, or it may be used as a general term for the different measures of the size of atoms, both bound in molecules and free.
EDTA	EDTA is a widely used acronym for the chemical compound ethylenediaminetetraacetic acid EDTA is a polyamino carboxylic acid with the formula $[CH_2N(CH_2CO_2H)_2]_2$. This colourless, water-soluble solid is produced on a large scale for many applications.
Nuclear reaction	In nuclear physics, a Nuclear reaction is the process in which two nuclei or nuclear particles collide to produce products different from the initial particles. In principle a reaction can involve more than three particles colliding, but because the probability of three or more nuclei to meet at the same time at the same place is much less than for two nuclei, such an event is exceptionally rare. While the transformation is spontaneous in the case of radioactive decay, it is initiated by a particle in the case of a Nuclear reaction.
Principal quantum number	In atomic physics, the Principal quantum number symbolized as n is the first of a set of quantum numbers (which includes: the Principal quantum number, the azimuthal quantum number, the magnetic quantum number, and the spin quantum number) of an atomic orbital. There are a set of quantum numbers associated with the energy states of the atom. The four quantum numbers n, â„", m, and s specify the complete and unique quantum state of a single electron in an atom called its wavefunction or orbital.
Atomic number	In chemistry and physics, the Atomic number is the number of protons found in the nucleus of an atom and therefore identical to the charge number of the nucleus. It is conventionally represented by the symbol Z. The Atomic number uniquely identifies a chemical element. In an atom of neutral charge, Atomic number is equal to the number of electrons.
Electron affinity	The Electron affinity, E_{ea}, of an atom or molecule is the amount of energy required to detach an electron from a singly charged negative ion, i.e., the energy change for the process

Chapter 8. Electron Configurations and Periodicity

Chapter 8. Electron Configurations and Periodicity

	$X^- \rightarrow X + e^-$ An equivalent definition is the energy released ($E_{initial} - E_{final}$) when an electron is attached to a neutral atom or molecule. All elements whose Electron affinity have been measured using modern methods have a positive Electron affinity, but older texts mistakenly report that some elements such as alkaline earth metals have negative E_{ea}, meaning they would repel electrons. This is not recognized by modern chemists.
Electronegativity	Electronegativity, symbol χ, is a chemical property that describes the ability of an atom (or, more rarely, a functional group) to attract electrons (or electron density) towards itself in a covalent bond. An atom"s Electronegativity is affected by both its atomic weight and the distance that its valence electrons reside from the charged nucleus. The higher the associated Electronegativity number, the more an element or compound attracts electrons towards it.
Ion	An Ion is an atom or molecule where the total number of electrons is not equal to the total number of protons, giving it a net positive or negative electrical charge. Since protons are positively charged and electrons are negatively charged, if there are more electrons than protons, the atom or molecule will be negatively charged. This is called an an Ion , from the Greek á¼€vÎ¯ , meaning "up".
Alkali	In chemistry, an Alkali is a basic, ionic salt of an Alkali metal or Alkali ne earth metal element. Alkali s are best known for being bases that dissolve in water. Bases are compounds with a pH greater than 7.
Amphoteric	In chemistry, an amphoteric substance is one that can react as either an acid or base. Many metals (such as zinc, tin, lead, aluminium, and beryllium) and most metalloids have amphoteric oxides or hydroxides. Another class of amphoteric substances are amphiprotic molecules which can either donate or accept a proton.
Indium	Indium is a chemical element with chemical symbol In and atomic number 49. This rare, soft, malleable and easily fusible post-transition metal is chemically similar to aluminium or gallium but more closely resembles zinc . Indium"s current primary application is to form transparent electrodes from Indium tin oxide in liquid crystal displays, and this use largely determines its global mining production.
Metallurgy	Metallurgy is a domain of materials science that studies the physical and chemical behavior of metallic elements, their intermetallic compounds, and their mixtures, which are called alloys. It is also the technology of metals: the way in which science is applied to their practical use. Metallurgy is commonly used in the craft of metalworking.

Chapter 8. Electron Configurations and Periodicity

Chapter 8. Electron Configurations and Periodicity

Thallium	Thallium is a chemical element with the symbol Tl and atomic number 81. This soft gray malleable poor metal resembles tin but discolors when exposed to air. Approximately 60-70% of Thallium production is used in the electronics industry, and the rest is used in the pharmaceutical industry and in glass manufacturing.
Latex	Latex is a name collectively given to a group of similar preparations consisting of stable dispersions of polymer microparticles in a liquid matrix (usually water.) Latexes may be natural or synthetic. Synthetic latexes are usually produced by emulsion polymerization using a variety of initiators, surfactants and monomers; the latter commonly include: - vinyl acetate - SBR (styrene-butadiene) - acrylates while more exotic formulations include allylic compounds, vinyl malate or VEOVAs. Latexes are used in coatings (e.g. Latex paint) and glues because they solidify by coalescence of the polymer particles as the water evaporates, and therefore can form films without releasing potentially toxic organic solvents in the environment.
Nitrogen	Nitrogen is a chemical element that has the symbol N and atomic number 7 and atomic mass 14.00674 u. Elemental Nitrogen is a colorless, odorless, tasteless and mostly inert diatomic gas at standard conditions, constituting 78% by volume of Earth"s atmosphere. Many industrially important compounds, such as ammonia, nitric acid, organic nitrates , and cyanides, contain Nitrogen.
Reactivity	Reactivity refers to the rate at which a chemical substance tends to undergo a chemical reaction in time. In pure compounds, Reactivity is regulated by the physical properties of the sample. For instance, grinding a sample to a higher specific surface area increases its Reactivity.
Solution	In chemistry, a Solution is a homogeneous mixture composed of two or more substances. In such a mixture, a solute is dissolved in another substance, known as a solvent. Gases may dissolve in liquids, for example, carbon dioxide or oxygen in water.
Antimony	Antimony) is a chemical element with the symbol Sb and atomic number 51. A metalloid, Antimony has four allotropic forms. The stable form of Antimony is a blue-white metalloid.
Arsenic	Arsenic is the chemical element that has the symbol As and atomic number 33. Arsenic was first documented by Albertus Magnus in 1250. Its atomic mass is 74.92.
Bismuth	Bismuth is a chemical element that has the symbol Bi and atomic number 83. This trivalent poor metal chemically resembles arsenic and antimony. Bismuth is heavy and brittle; it has a silvery white color with a pink tinge due to the surface oxide.

Chapter 8. Electron Configurations and Periodicity

Chapter 8. Electron Configurations and Periodicity

Boron oxide	Boron oxide is one of the oxides of boron. It is white, glassy, and solid, also known as diboron trioxide, formula B_2O_3. It is almost always found as the vitreous (amorphic) form; however, it can be crystallized after extensive annealing.
Chalcogens	The Chalcogens are the chemical elements in group 16 (old-style: VIB or VIA) of the periodic table. This group is sometimes known as the oxygen family. It consists of the elements oxygen (O), sulfur (S), selenium (Se), tellurium (Te), the radioactive element polonium (Po), and the synthetic element ununhexium (Uuh.)
Chlorin	In organic chemistry, a Chlorin is a large heterocyclic aromatic ring consisting, at the core, of three pyrroles and one pyrroline coupled through four methine linkages. Unlike a porphyrin, a Chlorin is therefore largely aromatic but not aromatic through the entire circumference of the ring. Magnesium-containing Chlorin s are called chlorophylls, and are the central photosensitive pigment in chloroplasts.
Chlorine	Chlorine . As the chloride ion, which is part of common salt and other compounds, it is abundant in nature and necessary to most forms of life, including humans. In its elemental form (Cl_2 or "di Chlorine ") under standard conditions, Chlorine is a powerful oxidant and is used in bleaching and disinfectants.
Fluorine	Fluorine is a chemical element, represented by the symbol F, and the atomic number 9. Fluorine forms a single bond with itself in elemental form, resulting in the diatomic F_2 molecule. F_2 is a supremely reactive, poisonous, pale, yellowish brown gas.
Halogen	The Halogen s or Halogen elements are a series of nonmetal elements from Group 17 IUPAC Style (formerly: VII, VIIA) of the periodic table, comprising fluorine, (F); chlorine, (Cl); bromine, (Br); iodine, (I); and astatine, (At.) The undiscovered element 117, temporarily named ununseptium, may also be a Halogen The group of Halogen s is the only periodic table group which contains elements in all three familiar states of matter at standard temperature and pressure.
Lead	Lead is a main-group element with symbol Pb and atomic number 82. Lead is a soft, malleable poor metal, also considered to be one of the heavy metals. Lead has a bluish-white color when freshly cut, but tarnishes to a dull grayish color when exposed to air.
Ozone	Ozone or trioxygen (O_3) is a triatomic molecule, consisting of three oxygen atoms. It is an allotrope of oxygen that is much less stable than the diatomic O_2. Ground-level Ozone is an air pollutant with harmful effects on the respiratory systems of animals.

Chapter 8. Electron Configurations and Periodicity

Chapter 8. Electron Configurations and Periodicity

Polonium	Polonium is a chemical element with the symbol Po and atomic number 84, discovered in 1898 by Marie and Pierre Curie. A rare and highly radioactive metalloid, Polonium is chemically similar to bismuth and tellurium, and it occurs in uranium ores. Polonium has been studied for possible use in heating spacecraft.
Selenium	Selenium is a chemical element with the atomic number 34, represented by the chemical symbol Se, an atomic mass of 78.96. It is a nonmetal, chemically related to sulfur and tellurium, and rarely occurs in its elemental state in nature. Isolated Selenium occurs in several different forms, the most stable of which is a dense purplish-gray semi-metal form that is structurally a trigonal polymer chain.
Silicon	Silicon is the most common metalloid. It is a chemical element, which has the symbol Si and atomic number 14. The atomic mass is 28.0855.
Sulfur	Sulfur or sulphur is the chemical element that has the atomic number 16. It is denoted with the symbol S. It is an abundant, multivalent non-metal. Sulfur in its native form is a yellow crystalline solid.
Tin	Tin is a chemical element with the symbol Sn and atomic number 50. It is a main group metal in group 14 of the periodic table. Tin shows chemical similarity to both neighboring group 14 elements, germanium and lead, like the two possible oxidation states +2 and +4.
White phosphorus	White phosphorus is a flare- and smoke-producing agent and an incendiary agent that is made from a common allotrope of the chemical element phosphorus. The main utility of White phosphorus munitions is to create smokescreens to mask movement from the enemy, or to mask his fire. In contrast to other smoke-causing munitions, White phosphorus burns quickly causing an instant bank of smoke.
Gray	The Gray is the SI unit of absorbed radiation dose due to ionizing radiation (for example, X-rays.) One Gray is the absorption of one joule of energy, in the form of ionizing radiation, by one kilogram of matter. $$1 \text{ Gy} = 1\,\frac{\text{J}}{\text{kg}} = 1\text{ m}^2 \cdot \text{s}^{-2}$$ For X-rays and gamma rays, these are the same units as the sievert (Sv.)
Molecular model	A Molecular model in this article, is a physical model that represents molecules and their processes. The creation of mathematical models of molecular properties and behaviour is Molecular model ling, and their graphical depiction is molecular graphics, but these topics are closely linked and each uses techniques from the others. In this article, Molecular model will primarily refer to systems containing more than one atom and where nuclear structure is neglected.

Chapter 8. Electron Configurations and Periodicity

Chapter 8. Electron Configurations and Periodicity

Oxoacid	An Oxoacid is an acid which contains oxygen. More specifically, it is an acid which: 1. contains oxygen; 2. contains at least one other element; 3. has at least one hydrogen atom bound to oxygen; and 4. forms an ion by the loss of one or more protons. The name oxyacid is sometimes used, although this is not recommended. Generally, Oxoacid s are simply polyatomic ions with a hydrogen cation. Under Lavoisier"s original theory, all acids contained oxygen, which was named from the Greek οξυς (acid, sharp) and γεινομαι (geinomai) (engender.)
Phosphorus	Phosphorus is the chemical element that has the symbol P and atomic number 15. A multivalent nonmetal of the nitrogen group, Phosphorus is commonly found in inorganic phosphate rocks. Elemental Phosphorus exists in two major forms - white Phosphorus and red Phosphorus.
Silver	Silver is a chemical element with the chemical symbol Ag and atomic number 47. A soft, white, lustrous transition metal, it has the highest electrical conductivity of any element and the highest thermal conductivity of any metal. The metal occurs naturally in its pure, free form, as an alloy with gold (electrum) and other metals, and in minerals such as argentite and chlorargyrite.
Astatine	Astatine is a radioactive chemical element with the symbol At and atomic number 85. It is the heaviest of the discovered halogens. Although Astatine is produced by radioactive decay in nature, due to its short half life it is found only in minute amounts.
Radon	Radon is a chemical element with symbol Rn and atomic number 86. Radon is a colorless, odorless, tasteless, naturally occurring, radioactive noble gas that is formed from the decay of radium. It is one of the heaviest substances that remains a gas under normal conditions and is considered to be a health hazard.

Chapter 8. Electron Configurations and Periodicity

Chapter 9. Ionic and Covalent Bonding

Chemical bond	A Chemical bond is the physical process responsible for the attractive interactions between atoms and molecules, and that which confers stability to diatomic and polyatomic chemical compounds. The explanation of the attractive forces is a complex area that is described by the laws of quantum electrodynamics. In practice, however, chemists usually rely on quantum theory or qualitative descriptions that are less rigorous but more easily explained to describe Chemical bond ing.
Complex	In chemistry, a Complex is a structure consisting of a central atom or molecule connected to surrounding atoms or molecules. Originally, a Complex implied a reversible association of molecules, atoms, or ions through weak chemical bonds. As applied to coordination chemistry, this meaning has evolved.
Covalent bond	A Covalent bond is a form of chemical bonding that is characterized by the sharing of pairs of electrons between atoms, or between atoms and other Covalent bond s. In short, attraction-to-repulsion stability that forms between atoms when they share electrons is known as Covalent bond ing. Covalent bond ing includes many kinds of interaction, including σ-bonding, π-bonding, metal to non-metal bonding, agostic interactions, and three-center two-electron bonds.
Ionic bond	An Ionic bond is a type of chemical bond that involves a metal and a non-metal ion (or polyatomic ions such as ammonium) through electrostatic attraction. In short, it is a bond formed by the attraction between two oppositely charged ions. The metal donates one or more electrons, forming a positively charged ion or cation with a stable electron configuration.
Salt	Salt is a dietary mineral composed primarily of sodium chloride that is essential for animal life, but toxic to most land plants. Salt flavor is one of the basic tastes, an important preservative and a popular food seasoning.
Sodium	Sodium is a metallic element with a symbol Na and atomic number 11. It is a soft, silvery-white, highly reactive metal and is a member of the alkali metals within "group 1" (formerly known as "group IA.") It has only one stable isotope, ^{23}Na.
Acid	An Acid is traditionally considered any chemical compound that, when dissolved in water, gives a solution with a hydrogen ion activity greater than in pure water, i.e. a pH less than 7.0. That approximates the modern definition of Johannes Nicolaus Brønsted and Martin Lowry, who independently defined an Acid as a compound which donates a hydrogen ion (H^+) to another compound (called a base.) Common examples include acetic Acid and sulfuric Acid (used in car batteries.)
Base	In chemistry, a Base is most commonly thought of as an aqueous substance that can accept hydrogen ions. A Base is also often referred to as an alkali if OH^- ions are involved. This refers to the Brønsted-Lowry theory of acids and bases.

Chapter 9. Ionic and Covalent Bonding

Chapter 9. Ionic and Covalent Bonding

Chlorin	In organic chemistry, a Chlorin is a large heterocyclic aromatic ring consisting, at the core, of three pyrroles and one pyrroline coupled through four methine linkages. Unlike a porphyrin, a Chlorin is therefore largely aromatic but not aromatic through the entire circumference of the ring. Magnesium-containing Chlorin s are called chlorophylls, and are the central photosensitive pigment in chloroplasts.
Chlorine	Chlorine . As the chloride ion, which is part of common salt and other compounds, it is abundant in nature and necessary to most forms of life, including humans. In its elemental form (Cl_2 or "di Chlorine ") under standard conditions, Chlorine is a powerful oxidant and is used in bleaching and disinfectants.
Graphite	Graphite " href="/wiki/Cleavage_(crystal)">loose interlamellar coupling between sheets in the structure, in fact in a vacuum environment (such as in technologies for use in space), Graphite was found to be a very poor lubricant. This fact led to the discovery that Graphite"s lubricity is due to adsorbed air and water between the layers, unlike other layered dry lubricants such as molybdenum disulfide. Recent studies suggest that an effect called superlubricity can also account for this effect.
Intermolecular forces	In physics, chemistry, and biology, Intermolecular forces are forces that act between stable molecules or between functional groups of macromolecules. Intermolecular forces include momentary attractions between molecules, diatomic free elements, and individual atoms. They differ from covalent and ionic bonding in that they are not stable, but are caused by momentary polarization of particles.
Ion	An Ion is an atom or molecule where the total number of electrons is not equal to the total number of protons, giving it a net positive or negative electrical charge. Since protons are positively charged and electrons are negatively charged, if there are more electrons than protons, the atom or molecule will be negatively charged. This is called an an Ion , from the Greek á¼€vÎ¯ , meaning "up".
Melting	Melting is a physical process that results in the phase change of a substance from a solid to a liquid. The internal energy of a solid substance is increased, typically by the application of heat or pressure, resulting in a rise of its temperature to the Melting point, at which the rigid ordering of molecular entities in the solid breaks down to a less-ordered state and the solid liquefies. An object that has melted completely is molten.
Melting point	The Melting point of a solid is the temperature range at which it changes state from solid to liquid. At the Melting point the solid and liquid phase exist in equilibrium. When considered as the temperature of the reverse change from liquid to solid, it is referred to as the freezing point.

Chapter 9. Ionic and Covalent Bonding

Chapter 9. Ionic and Covalent Bonding

Metallic bonding	Metallic bonding is the electromagnetic interaction between delocalized electrons, called conduction electrons, and the metallic nuclei within metals. Understood as the sharing of "free" electrons among a lattice of positively-charged ions (cations), Metallic bonding is sometimes compared with that of molten salts; however, this simplistic view holds true for very few metals. In a more quantum-mechanical view, the conduction electrons divide their density equally over all atoms that function as neutral (non-charged) entities.
Impurities	Impurities are substances inside a confined amount of liquid, gas which differ from the chemical composition of the material or compound. Impurities are either naturally occurring or added during synthesis of a chemical or commercial product. During production, Impurities may be purposely, accidentally, inevitably, or incidentally added into the substance.
Boron	Boron is the chemical element with atomic number 5 and the chemical symbol B. Boron is a trivalent metalloid element which occurs abundantly in the evaporite ores borax and ulexite. Several allotropes of Boron exist; amorphous Boron is a brown powder, though crystalline Boron is black, extremely hard, and a poor conductor at room temperature. Elemental Boron is used as a dopant in the semiconductor industry, while Boron compounds play important roles as light structural materials, insecticides and preservatives, and reagents for chemical synthesis.
Ionization	Ionization is the physical process of converting an atom or molecule into an ion by adding or removing charged particles such as electrons or other ions. This is often confused with dissociation (chemistry.) The process works slightly differently depending on whether an ion with a positive or a negative electric charge is being produced.
Fluorine	Fluorine is a chemical element, represented by the symbol F, and the atomic number 9. Fluorine forms a single bond with itself in elemental form, resulting in the diatomic F_2 molecule. F_2 is a supremely reactive, poisonous, pale, yellowish brown gas.
Lattice energy	The Lattice energy of an ionic solid is a measure of the strength of bonds in that ionic compound. It is usually defined as the enthalpy of formation of the ionic compound from gaseous ions and as such is invariably exothermic. $Na^+ (g) + Cl^- (g) \rightarrow NaCl (s)$ The experimental Lattice energy of NaCl is −787 kJ/mol.
Born-Haber cycle	The Born-Haber cycle is an approach to analyzing reaction energies. It was named after and developed by the two German scientists Max Born and Fritz Haber. The Born-Haber cycle involves the formation of an ionic compound from the reaction of a metal with a non-metal.

Chapter 9. Ionic and Covalent Bonding

Chapter 9. Ionic and Covalent Bonding

Dissociation	Dissociation in chemistry and biochemistry is a general process in which ionic compounds (complexes, molecules ions usually in a reversible manner. When a Bronsted-Lowry acid is put in water, a covalent bond between an electronegative atom and a hydrogen atom is broken by heterolytic fission, which gives a proton and a negative ion. Dissociation is the opposite of association and recombination.
Sublimation	Sublimation of an element or compound is a transition from the solid to gas phase with no intermediate liquid stage. Sublimation is an endothermic phase transition that occurs at temperatures and pressures below the triple point At normal pressures, most chemical compounds and elements possess three different states at different temperatures.
Ionic liquids	Ionic liquids, originally known as liquid electrolytes, ionic melts, ionic fluids, fused salts, liquid salts are liquids comprised predominantly of ions and ion-pairs at some given temperature. The term includes all classical molten salts, which are comprised of more thermally stable ions, such as sodium with chloride or potassium with nitrate, and has been attested as early as 1943. Recently, it has come to be used for salts whose melting point is relatively low (below 100 °C.)
Magnesium	Magnesium is a chemical element with the symbol Mg, atomic number 12, atomic weight 24.3050 and common oxidation number +2. Magnesium, an alkaline earth metal, is the ninth most abundant element in the universe by mass. The commonness of Magnesium is related to the fact that it is easily built up in supernova stars from a sequential addition of three helium nuclei to carbon .
Chemical formula	A Chemical formula is a way of expressing information about the atoms that constitute a particular chemical compound, and how the relationship between those atoms changes in chemical reactions. For molecular compounds it is also known as the molecular formula, and identifies each constituent element by its chemical symbol and indicates the number of atoms of each element found in each discrete molecule of that compound. If a molecule contains more than one atom of a particular element, this quantity is indicated using a subscript after the chemical symbol (although 19th-century books often used superscripts.)
Green chemistry	Green chemistry is a chemical philosophy encouraging the design of products and processes that reduce or eliminate the use and generation of hazardous substances. Whereas environmental chemistry is the chemistry of the natural environment, and of pollutant chemicals in nature, Green chemistry seeks to reduce and prevent pollution at its source. In 1990 the Pollution Prevention Act was passed in the United States.
Solvent	A Solvent is a liquid or gas that dissolves a solid, liquid resulting in a solution. The most common Solvent in everyday life is water. Most other commonly-used Solvent s are organic (carbon-containing) chemicals.

Chapter 9. Ionic and Covalent Bonding

Chapter 9. Ionic and Covalent Bonding

White phosphorus	White phosphorus is a flare- and smoke-producing agent and an incendiary agent that is made from a common allotrope of the chemical element phosphorus. The main utility of White phosphorus munitions is to create smokescreens to mask movement from the enemy, or to mask his fire. In contrast to other smoke-causing munitions, White phosphorus burns quickly causing an instant bank of smoke.
Acid rain	Acid rain is rain or any other form of precipitation that is unusually acidic. It has harmful effects on plants, aquatic animals, and infrastructure. Acid rain is mostly caused by human emissions of sulfur and nitrogen compounds which react in the atmosphere to produce acids.
Chemistry	In the history of science, the etymology of the word Chemistry is a debatable issue. It is agreed that the word "alchemy" is a European one, derived from Arabic, but the origin of the root word, chem, is uncertain. Words similar to it have been found in most ancient languages, with different meanings, but conceivably somehow related to alchemy.
Phosphorus	Phosphorus is the chemical element that has the symbol P and atomic number 15. A multivalent nonmetal of the nitrogen group, Phosphorus is commonly found in inorganic phosphate rocks. Elemental Phosphorus exists in two major forms - white Phosphorus and red Phosphorus.
Group	In chemistry, a Group is a vertical column in the periodic table of the chemical elements. The name family is derived from the fact that the elements share similar characteristics and traits, just as members of any human family would. There are 18 groups in the standard periodic table.
Monatomic ion	A Monatomic ion is an ion consisting of only one type of element atoms, unlike a polyatomic ion, which consists of more than one type of element in one ion. A type I binary ionic compound contains a metal (cation) that forms only one type of ion. A type II ionic compound contains a metal that forms more than one type of ion, i.e., ions with different charges *note that mercury (I) ions always occur bound together to form Hg_2.
Thallium	Thallium is a chemical element with the symbol Tl and atomic number 81. This soft gray malleable poor metal resembles tin but discolors when exposed to air. Approximately 60-70% of Thallium production is used in the electronics industry, and the rest is used in the pharmaceutical industry and in glass manufacturing.
Tin	Tin is a chemical element with the symbol Sn and atomic number 50. It is a main group metal in group 14 of the periodic table. Tin shows chemical similarity to both neighboring group 14 elements, germanium and lead, like the two possible oxidation states +2 and +4.

Chapter 9. Ionic and Covalent Bonding

Abundance	The abundance of a chemical element measures how relatively common the element is, or how much of the element there is by comparison to all other elements. abundance may be variously measured by the mass-fraction (the same as weight fraction), or mole-fraction (fraction of atoms, or sometimes fraction of molecules, in gases), or by volume fraction. Measurement by volume-fraction is a common abundance measure in mixed gases such as atmospheres, which is close to molecular mole-fraction for ideal gas mixtures (i.e., gas mixtures at relatively low densities and pressures.)
Carbon	Carbon is the chemical element with symbol C and atomic number 6. As a member of group 14 on the periodic table, it is nonmetallic and tetravalent--making four electrons available to form covalent chemical bonds. There are three naturally occurring isotopes, with ^{12}C and ^{13}C being stable, while ^{14}C is radioactive, decaying with a half-life of about 5730 years.
Configuration	The configuration of a molecule is the permanent geometry that results from the spatial arrangement of its bonds. The ability of the same set of atoms to form two or more molecules with different configurations is stereoisomerism. configuration is distinct from chemical conformation, a shape attainable by bond rotations.
Electron	The Electron is a subatomic particle that carries a negative electric charge. It has no known substructure and is believed to be a point particle. An Electron has a mass that is approximately 1836 times less than that of the proton.
Electron configuration	In atomic physics and quantum chemistry, Electron configuration is the arrangement of electrons of an atom, a molecule, or other physical structure. It concerns the way electrons can be distributed in the orbitals of the given system (atomic or molecular for instance.) Like other elementary particles, the electron is subject to the laws of quantum mechanics, and exhibits both particle-like and wave-like nature.
Ionic compound	In chemistry, an Ionic compound is a chemical compound in which ions are held together in a lattice structure by ionic bonds. Usually, the positively charged portion consists of metal cations and the negatively charged portion is an anion or polyatomic ion. Ions in Ionic compound s are held together by the electrostatic force between oppositely charged bodies.
Bismuth	Bismuth is a chemical element that has the symbol Bi and atomic number 83. This trivalent poor metal chemically resembles arsenic and antimony. Bismuth is heavy and brittle; it has a silvery white color with a pink tinge due to the surface oxide.
Polyaniline	Polyaniline is a conducting polymer of the semi-flexible rod polymer family. Although it was discovered over 150 years ago, only recently has Polyaniline captured the attention of the scientific community due to the discovery of its high electrical conductivity. Amongst the family of conducting polymers, Polyaniline is unique due to its ease of synthesis, environmental stability, and simple doping/dedoping chemistry.

Chapter 9. Ionic and Covalent Bonding

Chapter 9. Ionic and Covalent Bonding

Polyatomic ion	A Polyatomic ion is a charged species composed of two or more atoms covalently bonded or of a metal complex that can be considered as acting as a single unit in the context of acid and base chemistry or in the formation of salts. The prefix poly- means many in Greek, but even ions of two atoms are commonly referred to as polyatomic. In older literature, a Polyatomic ion is also referred to as a radical, and less commonly, as a radical group.
Electron affinity	The Electron affinity, E_{ea}, of an atom or molecule is the amount of energy required to detach an electron from a singly charged negative ion, i.e., the energy change for the process $$X^- \rightarrow X + e^-$$ An equivalent definition is the energy released ($E_{initial} - E_{final}$) when an electron is attached to a neutral atom or molecule. All elements whose Electron affinity have been measured using modern methods have a positive Electron affinity, but older texts mistakenly report that some elements such as alkaline earth metals have negative E_{ea}, meaning they would repel electrons. This is not recognized by modern chemists.
Iron	Iron is a chemical element with the symbol Fe and atomic number 26. Iron is a group 8 and period 4 element. Iron and Iron alloys (steels) are by far the most common metals and the most common ferromagnetic materials in everyday use.
Isomer	In chemistry, Isomer s are compounds with the same molecular formula but different structural formula. Isomer s do not necessarily share similar properties unless they also have the same functional groups. This should not be confused with a nuclear Isomer which involves a nucleus at different states of excitement.
Metal	In chemistry, a Metal is an element, compound, or alloy characterized by high electrical conductivity. In a Metal atoms readily lose electrons to form positive ions ; those ions are surrounded by delocalized electrons, which are responsible for the conductivity. The thus produced solid is held by electrostatic interactions between the ions and the electron cloud, which are called Metal lic bonds.
Lithium	Lithium is the chemical element with atomic number 3, and is represented by the symbol Li. It is a soft alkali metal with a silver-white color. Under standard conditions it is the lightest metal and the least dense solid element.
Molecular orbital	In chemistry, a Molecular orbital is a mathematical function that describes the wave-like behavior of an electron in a molecule. This function can be used to calculate chemical and physical properties such as the probability of finding an electron in any specific region. The use of the term "orbital" was first used in English by Robert S. Mulliken in 1925 as the English translation of Schrödinger"s use of the German word, "Eigenfunktion".

Chapter 9. Ionic and Covalent Bonding

Chapter 9. Ionic and Covalent Bonding

Crystal	A Crystal or Crystal line solid is a solid material whose constituent atoms, molecules, or ions are arranged in an orderly repeating pattern extending in all three spatial dimensions. The scientific study of Crystal s and Crystal formation is Crystal lography. The process of Crystal formation via mechanisms of Crystal growth is called Crystal lization or solidification.
Isoelectronic	Two or more molecular entities (atoms, molecules, ions) are described as being isoelectronic with each other if they have the same number of valence electrons and the same structure (number and connectivity of atoms), regardless of the nature of the elements involved.
	The N atom and the O^+ radical ion are isoelectronic because each has 5 electrons in the outer electronic shell. Similarly, the cations K^+, Ca^{2+}, and Sc^{3+}, the anions Cl^-, S^{2-}, and P^{3-} are all isoelectronic with the Ar atom.
Periodic trends	In Chemistry, Periodic trends are the tendencies of certain elemental characteristics to increase or decrease as one progresses from one corner of the Periodic table of elements.
	The atomic radius is the distance from the atomic nucleus to the outermost stable electron orbital in an atom that is at equilibrium. The atomic radius tends to decrease as one progresses across a period because the effective nuclear charge increases, thereby attracting the orbiting electrons and lessening the radius.
Ionic radius	Ionic radius, r_{ion}, is a measure of the size of an ion in a crystal lattice. It is measured in either picometres (pm) or Angstrom (Å), with 1 Å = 100 pm. Typical values range from 30 pm (0.3 Å) to over 200 pm (2 Å.)
Electron pair	In chemistry, an Electron pair consists of two electrons that occupy the same orbital but have opposite spins. MO diagrams depicting covalent (left) and polar covalent (right) bonding in a diatomic molecule. In both cases a bond is created by the formation of an Electron pair.
	Because electrons are fermions, the Pauli exclusion principle forbids these particles from having exactly the same quantum numbers.
Bond dissociation energy	In chemistry, Bond dissociation energy, D_0 or Bond dissociation energy, is one measure of the bond strength in a chemical bond. It is defined as the standard enthalpy change when a bond is cleaved by homolysis, with reactants and products of the homolysis reaction at 0 K (absolute zero.) For instance, the Bond dissociation energy for one of the C-H bonds in ethane (C_2H_6) is defined by the process:
	CH_3CH_2-H → Ethyl Radical + H.
	$D_0 = \Delta H$ = 101.1 kcal/mol (423.0 kJ/mol)
	The Bond dissociation energy is sometimes also called the bond dissociation enthalpy (or bond enthalpy), but these terms are not strictly equivalent, as they refer to the above reaction enthalpy at standard conditions, and differ from D_0 by about 1.5 kcal/mol (6 kJ/mol) in the case of a bond to hydrogen.

Chapter 9. Ionic and Covalent Bonding

Chapter 9. Ionic and Covalent Bonding

Bond energy	In chemistry, Bond energy is a measure of bond strength in a chemical bond. For example the carbon-hydrogen Bond energy in methane E(C-H) is the enthalpy change involved with breaking up one molecule of methane into a carbon atom and 4 hydrogen radicals divided by 4. Bond energy (E) should not be confused with bond dissociation energy.
Bond length	In molecular geometry, Bond length or bond distance is the average distance between nuclei of two bonded atoms in a molecule. Bond length is related to bond order, when more electrons participate in bond formation the bond will get shorter. Bond length is also inversely related to bond strength and the bond dissociation energy, as a stronger bond is also a shorter bond.
Hydrogen	Hydrogen is the chemical element with atomic number 1. It is represented by the symbol H. At standard temperature and pressure, Hydrogen is a colorless, odorless, nonmetallic, tasteless, highly flammable diatomic gas with the molecular formula H_2. With an atomic weight of 1.007 94 u, Hydrogen is the lightest element.
Lone pair	A Lone pair is a (valence) electron pair without bonding or sharing with other atoms. They are found in the outermost electron shell of an atom, so Lone pair s are a subset of a molecule"s valence electrons. They can be identified by examining the outermost energy level of an atom -- lone electron pairs consist of paired electrons as opposed to single electrons, which may appear if the atomic orbital is not full.
Octet rule	The Octet rule is a simple chemical rule of thumb that states that atoms tend to combine in such a way that they each have eight electrons in their valence shells, giving them the same electronic configuration as a noble gas. The rule is applicable to the main-group elements, especially carbon, nitrogen, oxygen, and the halogens, but also to metals such as sodium or magnesium. In simple terms, molecules or ions tend to be most stable when the outermost electron shells of their constituent atoms contain eight electrons.
Nitrogen	Nitrogen is a chemical element that has the symbol N and atomic number 7 and atomic mass 14.00674 u. Elemental Nitrogen is a colorless, odorless, tasteless and mostly inert diatomic gas at standard conditions, constituting 78% by volume of Earth"s atmosphere. Many industrially important compounds, such as ammonia, nitric acid, organic nitrates , and cyanides, contain Nitrogen.
Plutonium	Plutonium is a rare transuranic radioactive element. It is an actinide metal of silvery-white appearance that tarnishes when exposed to air, forming a dull coating when oxidized. The element normally exhibits six allotropes and four oxidation states.

Chapter 9. Ionic and Covalent Bonding

Chapter 9. Ionic and Covalent Bonding

Electronegativity	Electronegativity, symbol χ, is a chemical property that describes the ability of an atom (or, more rarely, a functional group) to attract electrons (or electron density) towards itself in a covalent bond. An atom"s Electronegativity is affected by both its atomic weight and the distance that its valence electrons reside from the charged nucleus. The higher the associated Electronegativity number, the more an element or compound attracts electrons towards it.
Polarity	In chemistry, polarity refers to a separation of electric charge leading to a molecule having an electric dipole. Polar molecules can bond together due to dipole-dipole intermolecular forces between one molecule (or part of a large molecule) with asymmetrical charge distribution and another molecule also with asymmetrical charge distribution. Molecular polarity is dependent on the difference in electronegativity between atoms in a compound and the asymmetry of the compound"s structure.
Activity	In chemical thermodynamics Activity is a measure of the "effective concentration" of a species in a mixture. By convention, it is a dimensionless quantity. The Activity of pure substances in condensed phases (solid or liquids) is normally taken as unity.
Molecule	A Molecule is defined as a sufficiently stable, electrically neutral group of at least two atoms in a definite arrangement held together by very strong (covalent) chemical bonds. Molecule s are distinguished from polyatomic ions in this strict sense. In organic chemistry and biochemistry, the term Molecule is used less strictly and also is applied to charged organic Molecule s and bio Molecule s.
Oxoacid	An Oxoacid is an acid which contains oxygen. More specifically, it is an acid which: 1. contains oxygen; 2. contains at least one other element; 3. has at least one hydrogen atom bound to oxygen; and 4. forms an ion by the loss of one or more protons. The name oxyacid is sometimes used, although this is not recommended. Generally, Oxoacid s are simply polyatomic ions with a hydrogen cation. Under Lavoisier"s original theory, all acids contained oxygen, which was named from the Greek οξυς (acid, sharp) and γεινομαι (geinomai) (engender.)
Formal charge	In chemistry, a Formal charge is a partial charge on an atom in a molecule assigned by assuming that electrons in a chemical bond are shared equally between atoms, regardless of relative electronegativity or in another definition the charge remaining on an atom when all ligands are removed homolytically . The Formal charge of any atom in a molecule can be calculated by the following equation: Formal charge = number of valence electrons of the atom in isolation - lone pair electrons on this atom in the molecule - half the total number of electrons participating in covalent bonds with this atom in the molecule.

Chapter 9. Ionic and Covalent Bonding

Chapter 9. Ionic and Covalent Bonding

	When determining the correct Lewis structure (or predominant resonance structure) for a molecule, the structure is chosen such that the Formal charge on each of the atoms is minimized.
Sulfur	Sulfur or sulphur is the chemical element that has the atomic number 16. It is denoted with the symbol S. It is an abundant, multivalent non-metal. Sulfur in its native form is a yellow crystalline solid.
Phenolphthalein	Phenolphthalein is a well-known chemical compound commonly pronounced [ËŒfi nÉ"l ËˆÎ¸ei lin]. It has the formula $C_{20}H_{14}O_4$ and is often written as "HIn" or "phph" in shorthand notation. Often used in titrations, it turns colorless in acidic solutions and pink in basic solutions.
Benzene	Benzene, or benzol, is an organic chemical compound and a known carcinogen with the molecular formula C_6H_6. It is sometimes abbreviated Ph-H. Benzene is a colorless and highly flammable liquid with a sweet smell and a relatively high melting point. Because it is a known carcinogen, its use as an additive in gasoline is now limited, but it is an important industrial solvent and precursor in the production of drugs, plastics, synthetic rubber, and dyes.
Resonance	Resonance in chemistry is a key component of valence bond theory used to graphically represent and mathematically model certain types of molecular structures when no single, conventional Lewis structure can satisfactorily represent the observed structure or explain its properties. Resonance instead considers such molecules to be an intermediate or average (called a Resonance hybrid) between several Lewis structures that differ only in the placement of the valence electrons. Scheme 1.
Ozone	Ozone or trioxygen (O_3) is a triatomic molecule, consisting of three oxygen atoms. It is an allotrope of oxygen that is much less stable than the diatomic O_2. Ground-level Ozone is an air pollutant with harmful effects on the respiratory systems of animals.
Ammonia	Ammonia is a compound of nitrogen and hydrogen with the formula NH_3. It is normally encountered as a gas with a characteristic pungent odor. Ammonia contributes significantly to the nutritional needs of terrestrial organisms by serving as a precursor to foodstuffs and fertilizers.
Aqueous solution	An Aqueous solution is a solution in which the solvent is water. It is usually shown in chemical equations by appending (aq) to the relevant formula. The word aqueous means pertaining to, related to, similar to, or dissolved in water.
Solution	In chemistry, a Solution is a homogeneous mixture composed of two or more substances. In such a mixture, a solute is dissolved in another substance, known as a solvent. Gases may dissolve in liquids, for example, carbon dioxide or oxygen in water.

Chapter 9. Ionic and Covalent Bonding

Chapter 9. Ionic and Covalent Bonding

Bond order	Bond order is the number of bonds between a pair of atoms. For example in nitrogen N≡N the Bond order is 3, in acetylene H−C≡C−H the Bond order between the two carbon atoms is also 3 and the C−H Bond order is 1. Bond order gives an indication to the stability of a bond.
Enthalpy	In thermodynamics and molecular chemistry, the Enthalpy is a thermodynamic property of a fluid. It can be used to calculate the heat transfer during a quasistatic process taking place in a closed thermodynamic system under constant pressure. Enthalpy H is an arbitrary concept but the Enthalpy change ΔH is more useful because it is equal to the change in the internal energy of the system, plus the work that the system has done on its surroundings.
Methanation	Methanation is a physical-chemical process to generate Methane from a mixture of various gases out of biomass fermentation or thermo-chemical gasification. The main components are carbon monoxide and hydrogen. The following main process describes the Methanation: $$CO + 3\,H_2 \rightarrow CH_4 + H_2O$$ This process is used for the generation of biogenous natural gas substitute, which can be fed into the gas grid.Methanation is the reverse reaction of steam methane reforming, which converts methane into synthesis gas.
Order	Order in a crystal lattice is the arrangement of some property with respect to atomic positions. It arises in charge ordering, spin ordering, magnetic ordering, and compositional ordering. It is a thermodynamic entropy concept often displayed by a second Order phase transition.
Infrared spectroscopy	Infrared spectroscopy is the subset of spectroscopy that deals with the infrared region of the electromagnetic spectrum. It covers a range of techniques, the most common being a form of absorption spectroscopy. As with all spectroscopic techniques, it can be used to identify compounds or investigate sample composition.
Polyethylene	Polyethylene or polythene (IUPAC name polyethene or poly(methylene)) is a thermoplastic commodity heavily used in consumer products (notably the plastic shopping bag.) Over 60 million tons of the material are produced worldwide every year. Polyethylene is a polymer consisting of long chains of the monomer ethylene (IUPAC name ethene.)
Spectroscopy	Spectroscopy was originally the study of the interaction between radiation and matter as a function of wavelength (λ.) In fact, historically, Spectroscopy referred to the use of visible light dispersed according to its wavelength, e.g. by a prism. Later the concept was expanded greatly to comprise any measurement of a quantity as function of either wavelength or frequency.

Chapter 9. Ionic and Covalent Bonding

Chapter 10. Molecular Geometry and Chemical Bonding Theory

Phosphorus	Phosphorus is the chemical element that has the symbol P and atomic number 15. A multivalent nonmetal of the nitrogen group, Phosphorus is commonly found in inorganic phosphate rocks. Elemental Phosphorus exists in two major forms - white Phosphorus and red Phosphorus.
Triclinic	In crystallography, the triclinic crystal system is one of the 7 lattice point groups. A crystal system is described by three basis vectors. In the triclinic system, the crystal is described by vectors of unequal length, as in the orthorhombic system.
Crystal	A Crystal or Crystal line solid is a solid material whose constituent atoms, molecules, or ions are arranged in an orderly repeating pattern extending in all three spatial dimensions. The scientific study of Crystal s and Crystal formation is Crystal lography. The process of Crystal formation via mechanisms of Crystal growth is called Crystal lization or solidification.
Crystal system	A Crystal system or crystal family or lattice system is one of six or seven categories of space groups, lattices, point groups, or crystals. Informally, two physical crystals tend to be in the same Crystal system if they have the same symmetries, though there are many exceptions to this. Unfortunately there are several slightly different conventions for the division into seven (or sometimes six) classes, differing mainly in the way crystals with exactly one axis of threefold rotation are classified, and the term Crystal system is used in at least 3 different ways in the literature.
Electron	The Electron is a subatomic particle that carries a negative electric charge. It has no known substructure and is believed to be a point particle. An Electron has a mass that is approximately 1836 times less than that of the proton.
Electron pair	In chemistry, an Electron pair consists of two electrons that occupy the same orbital but have opposite spins. MO diagrams depicting covalent (left) and polar covalent (right) bonding in a diatomic molecule. In both cases a bond is created by the formation of an Electron pair. Because electrons are fermions, the Pauli exclusion principle forbids these particles from having exactly the same quantum numbers.
Isomer	In chemistry, Isomer s are compounds with the same molecular formula but different structural formula. Isomer s do not necessarily share similar properties unless they also have the same functional groups. This should not be confused with a nuclear Isomer which involves a nucleus at different states of excitement.
Linear	In chemistry, the Linear molecular geometry describes the arrangement of three or more atoms placed at an expected bond angle of 180Å°. Linear organic molecules, e.g. acetylene, are often described by invoking sp orbital hybridization for the carbon centers. Many Linear molecules exist, prominent examples include CO_2, HCN, and xenon difluoride.
Molecular geometry	Molecular geometry or molecular structure is the three-dimensional arrangement of the atoms that constitute a molecule. It determines several properties of a substance including its reactivity, polarity, phase of matter, color, magnetism, and biological activity.

Chapter 10. Molecular Geometry and Chemical Bonding Theory

Chapter 10. Molecular Geometry and Chemical Bonding Theory

	The Molecular geometry can be determined by various spectroscopic methods and diffraction methods.
Covalent bond	A Covalent bond is a form of chemical bonding that is characterized by the sharing of pairs of electrons between atoms, or between atoms and other Covalent bond s. In short, attraction-to-repulsion stability that forms between atoms when they share electrons is known as Covalent bond ing. Covalent bond ing includes many kinds of interaction, including σ-bonding, π-bonding, metal to non-metal bonding, agostic interactions, and three-center two-electron bonds.
Beryllium	Beryllium is the chemical element with the symbol Be and atomic number 4. A bivalent element, Beryllium is found naturally only combined with other elements in minerals. Notable gemstones which contain Beryllium include Beryl (aquamarines and emeralds) and Chrysoberyl (Alexandrite and Cat"s eye.)
Lone pair	A Lone pair is a (valence) electron pair without bonding or sharing with other atoms. They are found in the outermost electron shell of an atom, so Lone pair s are a subset of a molecule"s valence electrons. They can be identified by examining the outermost energy level of an atom -- lone electron pairs consist of paired electrons as opposed to single electrons, which may appear if the atomic orbital is not full.
Atom	The Atom is a basic unit of matter consisting of a dense, central nucleus surrounded by a cloud of negatively charged electrons. The atomic nucleus contains a mix of positively charged protons and electrically neutral neutrons (except in the case of hydrogen-1, which is the only stable nuclide with no neutron.) The electrons of an Atom are bound to the nucleus by the electromagnetic force.
Boron	Boron is the chemical element with atomic number 5 and the chemical symbol B. Boron is a trivalent metalloid element which occurs abundantly in the evaporite ores borax and ulexite. Several allotropes of Boron exist; amorphous Boron is a brown powder, though crystalline Boron is black, extremely hard , and a poor conductor at room temperature. Elemental Boron is used as a dopant in the semiconductor industry, while Boron compounds play important roles as light structural materials, insecticides and preservatives, and reagents for chemical synthesis.
Carbon	Carbon is the chemical element with symbol C and atomic number 6. As a member of group 14 on the periodic table, it is nonmetallic and tetravalent--making four electrons available to form covalent chemical bonds. There are three naturally occurring isotopes, with ^{12}C and ^{13}C being stable, while ^{14}C is radioactive, decaying with a half-life of about 5730 years.
Rhombohedral	In crystallography, the rhombohedral crystal system is one of the seven lattice point groups the crystal is described by vectors of equal length, none of which are orthogonal. The rhombohedral system can be thought of as the cubic system stretched diagonally along a body. a = b = c; $\alpha, \beta, \gamma \neq 90°$.

Chapter 10. Molecular Geometry and Chemical Bonding Theory

Chapter 10. Molecular Geometry and Chemical Bonding Theory

Trigonal planar	In chemistry, trigonal planar is a molecular geometry with one atom at the center and three atoms at the corners of a triangle all in one plane. In an ideal trigonal planar species, all three ligands are identical and all bond angles are 120°. Such species belong to the point group D_{3h}.
Ammonia	Ammonia is a compound of nitrogen and hydrogen with the formula NH_3. It is normally encountered as a gas with a characteristic pungent odor. Ammonia contributes significantly to the nutritional needs of terrestrial organisms by serving as a precursor to foodstuffs and fertilizers.
Bent	In chemistry, the term "bent" can be applied to certain molecules to describe their molecular geometry. H_2O is an example of a bent molecule. The bond angle between the two hydrogen atoms is approximately 104.45°.
Octet rule	The Octet rule is a simple chemical rule of thumb that states that atoms tend to combine in such a way that they each have eight electrons in their valence shells, giving them the same electronic configuration as a noble gas. The rule is applicable to the main-group elements, especially carbon, nitrogen, oxygen, and the halogens, but also to metals such as sodium or magnesium. In simple terms, molecules or ions tend to be most stable when the outermost electron shells of their constituent atoms contain eight electrons.
Resonance	Resonance in chemistry is a key component of valence bond theory used to graphically represent and mathematically model certain types of molecular structures when no single, conventional Lewis structure can satisfactorily represent the observed structure or explain its properties. Resonance instead considers such molecules to be an intermediate or average (called a Resonance hybrid) between several Lewis structures that differ only in the placement of the valence electrons. Scheme 1.
Sulfur	Sulfur or sulphur is the chemical element that has the atomic number 16. It is denoted with the symbol S. It is an abundant, multivalent non-metal. Sulfur in its native form is a yellow crystalline solid.
Aqueous solution	An Aqueous solution is a solution in which the solvent is water. It is usually shown in chemical equations by appending (aq) to the relevant formula. The word aqueous means pertaining to, related to, similar to, or dissolved in water.
Methanation	Methanation is a physical-chemical process to generate Methane from a mixture of various gases out of biomass fermentation or thermo-chemical gasification. The main components are carbon monoxide and hydrogen. The following main process describes the Methanation: $$CO + 3\,H_2 \rightarrow CH_4 + H_2O$$ This process is used for the generation of biogenous natural gas substitute, which can be fed into the gas grid. Methanation is the reverse reaction of steam methane reforming, which converts methane into synthesis gas.

Chapter 10. Molecular Geometry and Chemical Bonding Theory

Chapter 10. Molecular Geometry and Chemical Bonding Theory

Solution	In chemistry, a Solution is a homogeneous mixture composed of two or more substances. In such a mixture, a solute is dissolved in another substance, known as a solvent. Gases may dissolve in liquids, for example, carbon dioxide or oxygen in water.
Nitrogen	Nitrogen is a chemical element that has the symbol N and atomic number 7 and atomic mass 14.00674 u. Elemental Nitrogen is a colorless, odorless, tasteless and mostly inert diatomic gas at standard conditions, constituting 78% by volume of Earth"s atmosphere. Many industrially important compounds, such as ammonia, nitric acid, organic nitrates , and cyanides, contain Nitrogen.
Ion	An Ion is an atom or molecule where the total number of electrons is not equal to the total number of protons, giving it a net positive or negative electrical charge. Since protons are positively charged and electrons are negatively charged, if there are more electrons than protons, the atom or molecule will be negatively charged. This is called an an Ion , from the Greek ἀνÎ¬ , meaning "up".
Chlorin	In organic chemistry, a Chlorin is a large heterocyclic aromatic ring consisting, at the core, of three pyrroles and one pyrroline coupled through four methine linkages. Unlike a porphyrin, a Chlorin is therefore largely aromatic but not aromatic through the entire circumference of the ring. Magnesium-containing Chlorin s are called chlorophylls, and are the central photosensitive pigment in chloroplasts.
Chlorine	Chlorine . As the chloride ion, which is part of common salt and other compounds, it is abundant in nature and necessary to most forms of life, including humans. In its elemental form (Cl_2 or "di Chlorine ") under standard conditions, Chlorine is a powerful oxidant and is used in bleaching and disinfectants.
Xenon difluoride	Xenon difluoride is a powerful fluorinating agent with the chemical formula XeF_2, and one of the most stable xenon compounds. Like most covalent inorganic fluorides it is moisture sensitive. It decomposes on contact with light or water vapour.
Dipole	In physics, there are two kinds of Dipole s : • An electric Dipole is a separation of positive and negative charge. The simplest example of this is a pair of electric charges of equal magnitude but opposite sign, separated by some, usually small, distance. A permanent electric Dipole is called an electret. • A magnetic Dipole is a closed circulation of electric current. A simple example of this is a single loop of wire with some constant current flowing through it.

Chapter 10. Molecular Geometry and Chemical Bonding Theory

Chapter 10. Molecular Geometry and Chemical Bonding Theory

	Dipole s can be characterized by their Dipole moment, a vector quantity. For the simple electric Dipole given above, the electric Dipole moment would point from the negative charge towards the positive charge, and have a magnitude equal to the strength of each charge times the separation between the charges. For the current loop, the magnetic Dipole moment would point through the loop (according to the right hand grip rule), with a magnitude equal to the current in the loop times the area of the loop. In addition to current loops, the electron, among other fundamental particles, is said to have a magnetic Dipole moment.
Iodine	Iodine , is a chemical element that has the symbol I and atomic number 53. Naturally-occurring Iodine is a single isotope with 74 neutrons. Chemically, Iodine is the second least reactive of the halogens, and the second most electropositive halogen; trailing behind astatine in both of these categories.
Square planar	The square planar molecular geometry in chemistry describes the stereochemistry (spatial arrangement of atoms) that is adopted by certain chemical compounds. As the name suggests, molecules of this geometry have their atoms positioned at the corners of a square on the same plane about a central atom. The addition of two ligands to linear compounds, ML_2, can afford square planar complexes.
Tellurium	Tellurium is a chemical element that has the symbol Te and atomic number 52. A brittle silver-white metalloid which looks like tin, Tellurium is chemically related to selenium and sulfur. Tellurium is primarily used in alloys and as a semiconductor.
Hydrogen	Hydrogen is the chemical element with atomic number 1. It is represented by the symbol H. At standard temperature and pressure, Hydrogen is a colorless, odorless, nonmetallic, tasteless, highly flammable diatomic gas with the molecular formula H_2. With an atomic weight of 1.007 94 u, Hydrogen is the lightest element.
Molecule	A Molecule is defined as a sufficiently stable, electrically neutral group of at least two atoms in a definite arrangement held together by very strong (covalent) chemical bonds. Molecule s are distinguished from polyatomic ions in this strict sense. In organic chemistry and biochemistry, the term Molecule is used less strictly and also is applied to charged organic Molecule s and bio Molecule s.
Acid	An Acid is traditionally considered any chemical compound that, when dissolved in water, gives a solution with a hydrogen ion activity greater than in pure water, i.e. a pH less than 7.0. That approximates the modern definition of Johannes Nicolaus Brønsted and Martin Lowry, who independently defined an Acid as a compound which donates a hydrogen ion (H^+) to another compound (called a base.) Common examples include acetic Acid and sulfuric Acid (used in car batteries.)

Chapter 10. Molecular Geometry and Chemical Bonding Theory

Chapter 10. Molecular Geometry and Chemical Bonding Theory

Water	Water is the most abundant molecule on Earth"s surface, constituting about 75% of the Earth"s surface. In nature it exists in liquid, solid, and gaseous states. It is in dynamic equilibrium between the liquid and gas states at standard temperature and pressure.
Boiling	Boiling, a type of phase transition, is the rapid vaporization of a liquid, which typically occurs when a liquid is heated to its Boiling point, the temperature at which the vapor pressure of the liquid is equal to the pressure exerted on the liquid by the surrounding environmental pressure. Thus, a liquid may also boil when the pressure of the surrounding atmosphere is sufficiently reduced, such as the use of a vacuum pump or at high altitudes. Boiling occurs in three characteristic stages, which are nucleate, transition and film Boiling.
Intermolecular forces	In physics, chemistry, and biology, Intermolecular forces are forces that act between stable molecules or between functional groups of macromolecules. Intermolecular forces include momentary attractions between molecules, diatomic free elements, and individual atoms. They differ from covalent and ionic bonding in that they are not stable, but are caused by momentary polarization of particles.
Atomic orbital	An Atomic orbital is a mathematical function that describes the wave-like behavior of either one electron or a pair of electrons, in an atom. This function can be used to calculate the probability of finding any electron of an atom in any specific region around the atom"s nucleus. These functions may serve as three-dimensional graph of an electron"s likely location.
Chemical bond	A Chemical bond is the physical process responsible for the attractive interactions between atoms and molecules, and that which confers stability to diatomic and polyatomic chemical compounds. The explanation of the attractive forces is a complex area that is described by the laws of quantum electrodynamics. In practice, however, chemists usually rely on quantum theory or qualitative descriptions that are less rigorous but more easily explained to describe Chemical bond ing.
Molecular orbital	In chemistry, a Molecular orbital is a mathematical function that describes the wave-like behavior of an electron in a molecule. This function can be used to calculate chemical and physical properties such as the probability of finding an electron in any specific region. The use of the term "orbital" was first used in English by Robert S. Mulliken in 1925 as the English translation of Schrödinger"s use of the German word, "Eigenfunktion".
Molecular orbital theory	In chemistry, Molecular orbital theory is a method for determining molecular structure in which electrons are not assigned to individual bonds between atoms, but are treated as moving under the influence of the nuclei in the whole molecule. In this theory, each molecule has a set of molecular orbitals, in which it is assumed that the molecular orbital wave function ψ_f may be written as a simple weighted sum of the n constituent atomic orbitals χ_i, according to the following equation: $$\psi_j = \sum_{i=1}^{n} c_{ij} \chi_i$$

Chapter 10. Molecular Geometry and Chemical Bonding Theory

Chapter 10. Molecular Geometry and Chemical Bonding Theory

	The c_{ij} coefficients may be determined numerically by substitution of this equation into the Schrödinger equation and application of the variational principle. This method is called the linear combination of atomic orbitals approximation and is used in computational chemistry.
Oxygen	Oxygen and -γενÎ®ς (-genÄ"s) (producer, literally begetter) is the element with atomic number 8 and represented by the symbol O. It is a member of the chalcogen group on the periodic table, and is a highly reactive nonmetallic period 2 element that readily forms compounds (notably oxides) with almost all other elements. At standard temperature and pressure two atoms of the element bind to form di Oxygen , a colorless, odorless, tasteless diatomic gas with the formula O_2. Oxygen is the third most abundant element in the universe by mass after hydrogen and helium and the most abundant element by mass in the Earth"s crust.
Valence	In chemistry, Valence is a measure of the number of chemical bonds formed by the atoms of a given element. Over the last century, the concept of Valence evolved into a range of approaches for describing the chemical bond, including Lewis structures, Valence bond theory, molecular orbitals, Valence shell electron pair repulsion theory and all the advanced methods of quantum chemistry.
	The etymology of the word "Valence" is from 1425, meaning "extract, preparation," from Latin valentia "strength, capacity," and the chemical meaning referring to the "combining power of an element" is recorded from 1884, from German Valenz.
Valence bond theory	In chemistry, Valence bond theory is one of two basic theories, along with molecular orbital theory, that developed to use the methods of quantum mechanics to explain chemical bonding. It focuses on how the atomic orbitals of the dissociated atoms combine on molecular formation to give individual chemical bonds. In contrast, molecular orbital theory has orbitals that cover the whole molecule.
Enthalpy	In thermodynamics and molecular chemistry, the Enthalpy is a thermodynamic property of a fluid. It can be used to calculate the heat transfer during a quasistatic process taking place in a closed thermodynamic system under constant pressure. Enthalpy H is an arbitrary concept but the Enthalpy change ΔH is more useful because it is equal to the change in the internal energy of the system, plus the work that the system has done on its surroundings.
Hydrocarbon	In organic chemistry, a Hydrocarbon is an organic compound consisting entirely of hydrogen and carbon. With relation to chemical terminology, aromatic Hydrocarbon s or arenes, alkanes, alkenes and alkyne-based compounds composed entirely of carbon or hydrogen are referred to as "pure" Hydrocarbon s, whereas other Hydrocarbon s with bonded compounds or impurities of sulfur or nitrogen, are referred to as "impure", and remain somewhat erroneously referred to as Hydrocarbon s.
	Hydrocarbon s are referred to as consisting of a "backbone" or "skeleton" composed entirely of carbon and hydrogen and other bonded compounds, and have a functional group that generally facilitates combustion.

Chapter 10. Molecular Geometry and Chemical Bonding Theory

Chapter 10. Molecular Geometry and Chemical Bonding Theory

Pi bond	In chemistry, Pi bond s (π bonds) are covalent chemical bonds where two lobes of one involved electron orbital overlap two lobes of the other involved electron orbital. Only one of the orbital"s nodal planes passes through both of the involved nuclei. Two p-orbitals forming a π-bond. The Greek letter π in their name refers to p orbitals, since the orbital symmetry of the Pi bond is the same as that of the p orbital when seen down the bond axis.
Sigma bonds	In chemistry, Sigma bonds are the strongest type of covalent chemical bond. Sigma bonding is most clearly defined for diatomic molecules using the language and tools of symmetry groups. In this formal approach, a σ-bond is symmetrical with respect to rotation about the bond axis.
Molecular model	A Molecular model in this article, is a physical model that represents molecules and their processes. The creation of mathematical models of molecular properties and behaviour is Molecular model ling, and their graphical depiction is molecular graphics, but these topics are closely linked and each uses techniques from the others. In this article, Molecular model will primarily refer to systems containing more than one atom and where nuclear structure is neglected.
Liquid oxygen	Liquid oxygen is a form of the element oxygen. It has a pale blue color and is strongly paramagnetic. Liquid oxygen has a density of 1.141 g/cmÂ³ (1.141 kg/L) and is moderately cryogenic (freezing point: 50.5 K (−222.65 °C), boiling point: 90.188 K (−182.96 °C) at 101.325 kPa (760 mm Hg.)
Antibonding	Antibonding is a type of chemical bonding. An Antibonding orbital is a form of molecular orbital (MO) that is located outside the region of two distinct nuclei. The overlap of the constituent atomic orbitals is said to be "out of phase" and as such the electrons present in each Antibonding orbital are repulsive and act to destabilize the molecule as a whole.
Bond order	Bond order is the number of bonds between a pair of atoms. For example in nitrogen N≡N the Bond order is 3, in acetylene H−C≡C−H the Bond order between the two carbon atoms is also 3 and the C−H Bond order is 1. Bond order gives an indication to the stability of a bond.
Helium	Helium is the chemical element with atomic number 2, and is represented by the symbol He. It is a colorless, odorless, tasteless, non-toxic, inert monatomic gas that heads the noble gas group in the periodic table. Its boiling and melting points are the lowest among the elements and it exists only as a gas except in extreme conditions.
Lithium	Lithium is the chemical element with atomic number 3, and is represented by the symbol Li. It is a soft alkali metal with a silver-white color. Under standard conditions it is the lightest metal and the least dense solid element.

Chapter 10. Molecular Geometry and Chemical Bonding Theory

Chapter 10. Molecular Geometry and Chemical Bonding Theory

Crystal structure	In mineralogy and crystallography, a Crystal structure is a unique arrangement of atoms in a crystal. A Crystal structure is composed of a motif, a set of atoms arranged in a particular way, and a lattice. Motifs are located upon the points of a lattice, which is an array of points repeating periodically in three dimensions.
Order	Order in a crystal lattice is the arrangement of some property with respect to atomic positions. It arises in charge ordering, spin ordering, magnetic ordering, and compositional ordering. It is a thermodynamic entropy concept often displayed by a second Order phase transition.
Diatomic molecules	Diatomic molecules are molecules composed only of two atoms, of either the same or different chemical elements. The prefix di- means two in Greek. Common Diatomic molecules are hydrogen, nitrogen, oxygen, and carbon monoxide.
Electron configuration	In atomic physics and quantum chemistry, Electron configuration is the arrangement of electrons of an atom, a molecule, or other physical structure. It concerns the way electrons can be distributed in the orbitals of the given system (atomic or molecular for instance.) Like other elementary particles, the electron is subject to the laws of quantum mechanics, and exhibits both particle-like and wave-like nature.
Configuration	The configuration of a molecule is the permanent geometry that results from the spatial arrangement of its bonds. The ability of the same set of atoms to form two or more molecules with different configurations is stereoisomerism. configuration is distinct from chemical conformation, a shape attainable by bond rotations.
Bond dissociation energy	In chemistry, Bond dissociation energy, D_0 or Bond dissociation energy, is one measure of the bond strength in a chemical bond. It is defined as the standard enthalpy change when a bond is cleaved by homolysis, with reactants and products of the homolysis reaction at 0 K (absolute zero.) For instance, the Bond dissociation energy for one of the C-H bonds in ethane (C_2H_6) is defined by the process: CH_3CH_2-H → Ethyl Radical + H. $D_0 = \Delta H = 101.1$ kcal/mol (423.0 kJ/mol) The Bond dissociation energy is sometimes also called the bond dissociation enthalpy (or bond enthalpy), but these terms are not strictly equivalent, as they refer to the above reaction enthalpy at standard conditions, and differ from D_0 by about 1.5 kcal/mol (6 kJ/mol) in the case of a bond to hydrogen.
Bond length	In molecular geometry, Bond length or bond distance is the average distance between nuclei of two bonded atoms in a molecule. Bond length is related to bond order, when more electrons participate in bond formation the bond will get shorter. Bond length is also inversely related to bond strength and the bond dissociation energy, as a stronger bond is also a shorter bond.

Chapter 10. Molecular Geometry and Chemical Bonding Theory

Chapter 10. Molecular Geometry and Chemical Bonding Theory

Pauli exclusion principle	The Pauli exclusion principle is a quantum mechanical principle formulated by Wolfgang Pauli in 1925. It states that no two identical fermions may occupy the same quantum state simultaneously. A more rigorous statement of this principle is that, for two identical fermions, the total wave function is anti-symmetric.
Dissociation	Dissociation in chemistry and biochemistry is a general process in which ionic compounds (complexes, molecules ions usually in a reversible manner. When a Bronsted-Lowry acid is put in water, a covalent bond between an electronegative atom and a hydrogen atom is broken by heterolytic fission, which gives a proton and a negative ion. Dissociation is the opposite of association and recombination.
Principle	A Principle is a comprehensive and fundamental law, doctrine, or assumption. It can be a rule or code of conduct. It can be a law or fact of nature underlying the working of an artificial device.
Group	In chemistry, a Group is a vertical column in the periodic table of the chemical elements. The name family is derived from the fact that the elements share similar characteristics and traits, just as members of any human family would. There are 18 groups in the standard periodic table.
Ozone	Ozone or trioxygen (O_3) is a triatomic molecule, consisting of three oxygen atoms. It is an allotrope of oxygen that is much less stable than the diatomic O_2. Ground-level Ozone is an air pollutant with harmful effects on the respiratory systems of animals.
Benzene	Benzene, or benzol, is an organic chemical compound and a known carcinogen with the molecular formula C_6H_6. It is sometimes abbreviated Ph-H. Benzene is a colorless and highly flammable liquid with a sweet smell and a relatively high melting point. Because it is a known carcinogen, its use as an additive in gasoline is now limited, but it is an important industrial solvent and precursor in the production of drugs, plastics, synthetic rubber, and dyes.
Acid rain	Acid rain is rain or any other form of precipitation that is unusually acidic. It has harmful effects on plants, aquatic animals, and infrastructure. Acid rain is mostly caused by human emissions of sulfur and nitrogen compounds which react in the atmosphere to produce acids.
Impurities	Impurities are substances inside a confined amount of liquid, gas which differ from the chemical composition of the material or compound. Impurities are either naturally occurring or added during synthesis of a chemical or commercial product. During production, Impurities may be purposely, accidentally, inevitably, or incidentally added into the substance.

Chapter 10. Molecular Geometry and Chemical Bonding Theory

Chapter 11. States of Matter; Liquids and Solids

Carbon	Carbon is the chemical element with symbol C and atomic number 6. As a member of group 14 on the periodic table, it is nonmetallic and tetravalent--making four electrons available to form covalent chemical bonds. There are three naturally occurring isotopes, with ^{12}C and ^{13}C being stable, while ^{14}C is radioactive, decaying with a half-life of about 5730 years.
Impurities	Impurities are substances inside a confined amount of liquid, gas which differ from the chemical composition of the material or compound. Impurities are either naturally occurring or added during synthesis of a chemical or commercial product. During production, Impurities may be purposely, accidentally, inevitably, or incidentally added into the substance.
Phase transition	In thermodynamics, a Phase transition is the transformation of a thermodynamic system from one phase to another. At a Phase transition point, physical properties may undergo abrupt change: for instance, the volume of the two phases may be vastly different as is illustrated by the boiling of liquid water to form steam. The term is most commonly used to describe transitions between solid, liquid and gaseous states of matter, in rare cases including plasma.
Vanadium	Vanadium is the chemical element with the symbol V and atomic number 23. It is a soft, silvery grey, ductile transition metal. The formation of an oxide layer stabilizes the metal against oxidation.
Covalent radius	The Covalent radius, r_{cov}, is a measure of the size of atom that forms part of a covalent bond. It is measured either in picometres (pm) or ångströms (Å), with 1 Å = 100 pm. In principle, the sum of the two covalent radii should equal the covalent bond length between two atoms.
Condensation	Condensation is the change of the physical state of aggregation (or simply state) of matter from gaseous phase into liquid phase. When the transition happens from the gaseous phase into the solid phase directly, bypassing the liquid phase, the change is called deposition. Condensation commonly occurs when a vapor is cooled to its dew point, but the dew point can also be reached through compression.
Freezing	In physical science, Freezing or solidification is the process in which a liquid turns into a solid when cold enough. The Freezing point is the temperature at which this happens. Melting, the process of turning a solid to a liquid, is almost the exact opposite of Freezing.
Melting	Melting is a physical process that results in the phase change of a substance from a solid to a liquid. The internal energy of a solid substance is increased, typically by the application of heat or pressure, resulting in a rise of its temperature to the Melting point, at which the rigid ordering of molecular entities in the solid breaks down to a less-ordered state and the solid liquefies. An object that has melted completely is molten.

Chapter 11. States of Matter; Liquids and Solids

Chapter 11. States of Matter; Liquids and Solids

Metal	In chemistry, a Metal is an element, compound, or alloy characterized by high electrical conductivity. In a Metal atoms readily lose electrons to form positive ions ; those ions are surrounded by delocalized electrons, which are responsible for the conductivity. The thus produced solid is held by electrostatic interactions between the ions and the electron cloud, which are called Metal lic bonds.
Sublimation	Sublimation of an element or compound is a transition from the solid to gas phase with no intermediate liquid stage. Sublimation is an endothermic phase transition that occurs at temperatures and pressures below the triple point At normal pressures, most chemical compounds and elements possess three different states at different temperatures.
Vapor	A Vapor or vapour is a substance in the gas phase at a temperature lower than its critical temperature. This means that the Vapor can be condensed to a liquid or to a solid by increasing its pressure, without reducing the temperature. For example, water has a critical temperature of 374°C (or 647 K) which is the highest temperature at which liquid water can exist.
Vapor pressure	Vapor pressure, is the pressure of a vapor in equilibrium with its non-vapor phases. All liquids and solids have a tendency to evaporate to a gaseous form, and all gases have a tendency to condense back into their original form At any given temperature, for a particular substance, there is a pressure at which the gas of that substance is in dynamic equilibrium with its liquid or solid forms.
Water	Water is the most abundant molecule on Earth"s surface, constituting about 75% of the Earth"s surface. In nature it exists in liquid, solid, and gaseous states. It is in dynamic equilibrium between the liquid and gas states at standard temperature and pressure.
Molecule	A Molecule is defined as a sufficiently stable, electrically neutral group of at least two atoms in a definite arrangement held together by very strong (covalent) chemical bonds. Molecule s are distinguished from polyatomic ions in this strict sense. In organic chemistry and biochemistry, the term Molecule is used less strictly and also is applied to charged organic Molecule s and bio Molecule s.
Boiling	Boiling, a type of phase transition, is the rapid vaporization of a liquid, which typically occurs when a liquid is heated to its Boiling point, the temperature at which the vapor pressure of the liquid is equal to the pressure exerted on the liquid by the surrounding environmental pressure. Thus, a liquid may also boil when the pressure of the surrounding atmosphere is sufficiently reduced, such as the use of a vacuum pump or at high altitudes. Boiling occurs in three characteristic stages, which are nucleate, transition and film Boiling.
Melting point	The Melting point of a solid is the temperature range at which it changes state from solid to liquid. At the Melting point the solid and liquid phase exist in equilibrium. When considered as the temperature of the reverse change from liquid to solid, it is referred to as the freezing point.

Chapter 11. States of Matter; Liquids and Solids

Chapter 11. States of Matter; Liquids and Solids

Volatility	Volatility in the context of chemistry, physics and thermodynamics is a measure of the tendency of a substance to vaporize. It has also been defined as a measure of how readily a substance vaporizes. At a given temperature, substances with higher vapor pressures will vaporize more readily than substances with a lower vapor pressure.
Chlorination	Chlorination is the process of adding the element chlorine to water as a method of water purification to make it fit for human consumption as drinking water. Water which has been treated with chlorine is effective in preventing the spread of disease. The Chlorination of public drinking supplies was originally met with resistance, as people were concerned about the health effects of the practice.
Ammonia	Ammonia is a compound of nitrogen and hydrogen with the formula NH_3. It is normally encountered as a gas with a characteristic pungent odor. Ammonia contributes significantly to the nutritional needs of terrestrial organisms by serving as a precursor to foodstuffs and fertilizers.
Enthalpy	In thermodynamics and molecular chemistry, the Enthalpy is a thermodynamic property of a fluid. It can be used to calculate the heat transfer during a quasistatic process taking place in a closed thermodynamic system under constant pressure. Enthalpy H is an arbitrary concept but the Enthalpy change ΔH is more useful because it is equal to the change in the internal energy of the system, plus the work that the system has done on its surroundings.
Enthalpy of fusion	The standard Enthalpy of fusion, also known as the heat of fusion or specific melting heat, is the amount of thermal energy which must be absorbed or evolved for 1 mole of a substance to change states from a solid to a liquid or vice versa. It is also called the latent heat of fusion or the enthalpy change of fusion, and the temperature at which it occurs is called the melting point. When thermal energy is withdrawn from a liquid or solid, the temperature falls.
Aqueous solution	An Aqueous solution is a solution in which the solvent is water. It is usually shown in chemical equations by appending (aq) to the relevant formula. The word aqueous means pertaining to, related to, similar to, or dissolved in water.
Solution	In chemistry, a Solution is a homogeneous mixture composed of two or more substances. In such a mixture, a solute is dissolved in another substance, known as a solvent. Gases may dissolve in liquids, for example, carbon dioxide or oxygen in water.
Phase diagram	A Phase diagram in physical chemistry, engineering, mineralogy, and materials science is a type of chart used to show conditions at which thermodynamically-distinct phases can occur at equilibrium. In mathematics and physics, "Phase diagram" is used with a different meaning: a synonym for a phase space. Common components of a Phase diagram are lines of equilibrium or phase boundaries, which refer to lines that mark conditions under which multiple phases can coexist at equilibrium.

Chapter 11. States of Matter; Liquids and Solids

Chapter 11. States of Matter; Liquids and Solids

Sulfur	Sulfur or sulphur is the chemical element that has the atomic number 16. It is denoted with the symbol S. It is an abundant, multivalent non-metal. Sulfur in its native form is a yellow crystalline solid.
Triple point	In thermodynamics, the Triple point of a substance is the temperature and pressure at which three phases (for example, gas, liquid, and solid) of that substance coexist in thermodynamic equilibrium. For example, the Triple point of mercury occurs at a temperature of −38.8344 °C and a pressure of 0.2 mPa. In addition to the Triple point between solid, liquid, and gas, there can be Triple point s involving more than one solid phase, for substances with multiple polymorphs.
Nitrogen	Nitrogen is a chemical element that has the symbol N and atomic number 7 and atomic mass 14.00674 u. Elemental Nitrogen is a colorless, odorless, tasteless and mostly inert diatomic gas at standard conditions, constituting 78% by volume of Earth"s atmosphere. Many industrially important compounds, such as ammonia, nitric acid, organic nitrates , and cyanides, contain Nitrogen.
Cadmium	Cadmium is a chemical element with the symbol Cd and atomic number 48. The soft, bluish-white transition metal is chemically similar to the two other metals in group 12 , zinc and mercury. Similar to zinc it prefers oxidation state +2 in most of its compounds and similar to mercury it shows a low melting point for a transition metal.
Solvent	A Solvent is a liquid or gas that dissolves a solid, liquid resulting in a solution. The most common Solvent in everyday life is water. Most other commonly-used Solvent s are organic (carbon-containing) chemicals.
Ion	An Ion is an atom or molecule where the total number of electrons is not equal to the total number of protons, giving it a net positive or negative electrical charge. Since protons are positively charged and electrons are negatively charged, if there are more electrons than protons, the atom or molecule will be negatively charged. This is called an an Ion , from the Greek á¼€νÎ¬ , meaning "up".
Soap	Soap is an anionic surfactant used in conjunction with water for washing and cleaning, which historically comes either in solid bars or in the form of a viscous liquid. Soap consists of sodium or potassium salts of fatty acids and is obtained by reacting common oils or fats with a strong alkaline solution (the base, popularly referred to as lye) in a process known as saponification. The fats are hydrolyzed by the base, yielding alkali salts of fatty acids (crude Soap) and glycerol.

Chapter 11. States of Matter; Liquids and Solids

Chapter 11. States of Matter; Liquids and Solids

Colloid	A Colloid is a type of chemical mixture where one substance is dispersed evenly throughout another. The particles of the dispersed substance are only suspended in the mixture, unlike a solution, where they are completely dissolved within. This occurs because the particles in a Colloid are larger than in a solution - small enough to be dispersed evenly and maintain a homogenous appearance, but large enough to scatter light and not dissolve.
Bromine	Bromine , Greek: βρά¿¶μος, brómos, meaning "stench "), is a chemical element with the symbol Br and atomic number 35. A halogen element, Bromine is a reddish-brown volatile liquid at standard room temperature that is intermediate in reactivity between chlorine and iodine. Bromine vapours are corrosive and toxic.
Neon	Neon is the chemical element that has the symbol Ne and atomic number 10. Although a very common element in the universe, it is rare on Earth. A colorless, inert noble gas under standard conditions, Neon gives a distinct reddish-orange glow when used in discharge tubes and Neon lamps.
Atomic mass	The Atomic mass is the mass of an atom, most often expressed in unified Atomic mass units. The Atomic mass may be considered to be the total mass of protons, neutrons and electrons in a single atom (when the atom is motionless.) The Atomic mass is sometimes incorrectly used as a synonym of relative Atomic mass, average Atomic mass and atomic weight; however, these differ subtly from the Atomic mass.
Configuration	The configuration of a molecule is the permanent geometry that results from the spatial arrangement of its bonds. The ability of the same set of atoms to form two or more molecules with different configurations is stereoisomerism. configuration is distinct from chemical conformation, a shape attainable by bond rotations.
Electron	The Electron is a subatomic particle that carries a negative electric charge. It has no known substructure and is believed to be a point particle. An Electron has a mass that is approximately 1836 times less than that of the proton.
Electron configuration	In atomic physics and quantum chemistry, Electron configuration is the arrangement of electrons of an atom, a molecule, or other physical structure. It concerns the way electrons can be distributed in the orbitals of the given system (atomic or molecular for instance.) Like other elementary particles, the electron is subject to the laws of quantum mechanics, and exhibits both particle-like and wave-like nature.
Intermolecular forces	In physics, chemistry, and biology, Intermolecular forces are forces that act between stable molecules or between functional groups of macromolecules. Intermolecular forces include momentary attractions between molecules, diatomic free elements, and individual atoms. They differ from covalent and ionic bonding in that they are not stable, but are caused by momentary polarization of particles.

Chapter 11. States of Matter; Liquids and Solids

Chapter 11. States of Matter; Liquids and Solids

Hydrogen	Hydrogen is the chemical element with atomic number 1. It is represented by the symbol H. At standard temperature and pressure, Hydrogen is a colorless, odorless, nonmetallic, tasteless, highly flammable diatomic gas with the molecular formula H_2. With an atomic weight of 1.007 94 u, Hydrogen is the lightest element.
Van der Waals force	In physical chemistry, the Van der Waals force (or van der Waals interaction) is the attractive or repulsive force between molecules (or between parts of the same molecule) other than those due to covalent bonds or to the electrostatic interaction of ions with one another or with neutral molecules. The term includes: • permanent dipole-permanent dipole forces • induced dipole-induced dipole forces • instantaneous induced dipole-induced dipole (London dispersion forces.) It is also sometimes used loosely as a synonym for the totality of intermolecular forces. Van der Waals force s are relatively weak compared to normal chemical bonds, but play a fundamental role in fields as diverse as supramolecular chemistry, structural biology, polymer science, nanotechnology, surface science, and condensed matter physics. Van der Waals force s define the chemical character of many organic compounds.
Chemical bond	A Chemical bond is the physical process responsible for the attractive interactions between atoms and molecules, and that which confers stability to diatomic and polyatomic chemical compounds. The explanation of the attractive forces is a complex area that is described by the laws of quantum electrodynamics. In practice, however, chemists usually rely on quantum theory or qualitative descriptions that are less rigorous but more easily explained to describe Chemical bond ing.
Acid	An Acid is traditionally considered any chemical compound that, when dissolved in water, gives a solution with a hydrogen ion activity greater than in pure water, i.e. a pH less than 7.0. That approximates the modern definition of Johannes Nicolaus Brønsted and Martin Lowry, who independently defined an Acid as a compound which donates a hydrogen ion (H^+) to another compound (called a base.) Common examples include acetic Acid and sulfuric Acid (used in car batteries.)
Methanation	Methanation is a physical-chemical process to generate Methane from a mixture of various gases out of biomass fermentation or thermo-chemical gasification. The main components are carbon monoxide and hydrogen. The following main process describes the Methanation: $$CO + 3\,H_2 \rightarrow CH_4 + H_2O$$ This process is used for the generation of biogenous natural gas substitute, which can be fed into the gas grid.Methanation is the reverse reaction of steam methane reforming, which converts methane into synthesis gas.

Chapter 11. States of Matter; Liquids and Solids

Chapter 11. States of Matter; Liquids and Solids

Complex	In chemistry, a Complex is a structure consisting of a central atom or molecule connected to surrounding atoms or molecules. Originally, a Complex implied a reversible association of molecules, atoms, or ions through weak chemical bonds. As applied to coordination chemistry, this meaning has evolved.
Diamond	In mineralogy, Diamond is an allotrope of carbon, where the carbon atoms are arranged in a variation of the face centered cubic crystal structure called a Diamond lattice. Diamond is the second most stable form of carbon, after graphite; however, the conversion rate from Diamond to graphite is negligible at ambient conditions. Diamond is specifically renowned as a material with superlative physical qualities, most of which originate from the strong covalent bonding between its atoms.
Group	In chemistry, a Group is a vertical column in the periodic table of the chemical elements. The name family is derived from the fact that the elements share similar characteristics and traits, just as members of any human family would. There are 18 groups in the standard periodic table.
Lattice energy	The Lattice energy of an ionic solid is a measure of the strength of bonds in that ionic compound. It is usually defined as the enthalpy of formation of the ionic compound from gaseous ions and as such is invariably exothermic. $Na^+ (g) + Cl^- (g) \rightarrow NaCl (s)$ The experimental Lattice energy of NaCl is −787 kJ/mol.
Magnesium	Magnesium is a chemical element with the symbol Mg, atomic number 12, atomic weight 24.3050 and common oxidation number +2. Magnesium, an alkaline earth metal, is the ninth most abundant element in the universe by mass. The commonness of Magnesium is related to the fact that it is easily built up in supernova stars from a sequential addition of three helium nuclei to carbon .
Silicon	Silicon is the most common metalloid. It is a chemical element, which has the symbol Si and atomic number 14. The atomic mass is 28.0855.
Sodium	Sodium is a metallic element with a symbol Na and atomic number 11. It is a soft, silvery-white, highly reactive metal and is a member of the alkali metals within "group 1" (formerly known as "group IA".) It has only one stable isotope, ^{23}Na.
Tungsten	Tungsten is a chemical element with the chemical symbol W and atomic number 74. A steel-gray metal, Tungsten is found in several ores, including wolframite and scheelite. It is remarkable for its robust physical properties, especially the fact that it has the highest melting point of all the non-alloyed metals and the second highest of all the elements after carbon.

Chapter 11. States of Matter; Liquids and Solids

Chapter 11. States of Matter; Liquids and Solids

Crystal	A Crystal or Crystal line solid is a solid material whose constituent atoms, molecules, or ions are arranged in an orderly repeating pattern extending in all three spatial dimensions. The scientific study of Crystal s and Crystal formation is Crystal lography. The process of Crystal formation via mechanisms of Crystal growth is called Crystal lization or solidification.
Crystal structure	In mineralogy and crystallography, a Crystal structure is a unique arrangement of atoms in a crystal. A Crystal structure is composed of a motif, a set of atoms arranged in a particular way, and a lattice. Motifs are located upon the points of a lattice, which is an array of points repeating periodically in three dimensions.
Hardness	Hardness refers to various properties of matter in the solid phase that give it high resistance to various kinds of shape change when force is applied. Hard matter is contrasted with soft matter. Macroscopic Hardness is generally characterized by strong intermolecular bonds.
Lithium	Lithium is the chemical element with atomic number 3, and is represented by the symbol Li. It is a soft alkali metal with a silver-white color. Under standard conditions it is the lightest metal and the least dense solid element.
Carbide	In chemistry, a Carbide is a compound composed of carbon and a less electronegative element. Carbide s can be generally classified by chemical bonding type as follows: (i) salt-like, (ii) covalent compounds, (iii) interstitial compounds, and (iv) "intermediate" transition metal Carbide s. Examples include calcium Carbide silicon Carbide tungsten Carbide (often called simply Carbide , and cementite, each used in key industrial applications.
Graphite	Graphite " href="/wiki/Cleavage_(crystal)">loose interlamellar coupling between sheets in the structure, in fact in a vacuum environment (such as in technologies for use in space), Graphite was found to be a very poor lubricant. This fact led to the discovery that Graphite"s lubricity is due to adsorbed air and water between the layers, unlike other layered dry lubricants such as molybdenum disulfide. Recent studies suggest that an effect called superlubricity can also account for this effect.
Copper	Copper is a chemical element with the symbol Cu and atomic number 29. It is a ductile metal with very high thermal and electrical conductivity. Pure Copper is rather soft and malleable and a freshly-exposed surface has a pinkish or peachy color.
Sucrose	Sucrose is a disaccharide of glucose and fructose with an α 1,2 glycosidic linkage. The molecular formula of Sucrose is $C_{12}H_{22}O_{11}$. Its systematic name is β-D-fructofuranosyl--α-D-glucopyranoside
Cubic	The cubic crystal system is a crystal system where the unit cell is in the shape of a cube. This is one of the most common and simplest shapes found in crystals and minerals. There are three main varieties of these crystals, called "simple cubic", "body-centered cubic" (bcc), and "face-centered cubic" (fcc), plus a number of other variants listed below.

Chapter 11. States of Matter; Liquids and Solids

Chapter 11. States of Matter; Liquids and Solids

Hexagonal	In crystallography, the hexagonal system is one of the 7 crystal systems. It contains 7 point groups. It has the same symmetry as a right prism with a hexagonal base.
Monoclinic	In crystallography, the monoclinic crystal system is one of the 7 lattice point groups. A crystal system is described by three vectors. In the monoclinic system, the crystal is described by vectors of unequal length, as in the orthorhombic system.
Orthorhombic	In crystallography, the orthorhombic crystal system is one of the seven lattice point groups. orthorhombic lattices result from stretching a cubic lattice along two of its orthogonal pairs by two different factors, resulting in a rectangular prism with a rectangular base (a by b) and height (c), such that a, b, and c are distinct. All three bases intersect at 90° angles.
Rhombohedral	In crystallography, the rhombohedral crystal system is one of the seven lattice point groups the crystal is described by vectors of equal length, none of which are orthogonal. The rhombohedral system can be thought of as the cubic system stretched diagonally along a body. a = b = c; $\alpha, \beta, \gamma \neq 90°$.
Tetragonal	In crystallography, the tetragonal crystal system is one of the 7 lattice point groups. tetragonal crystal lattices result from stretching a cubic lattice along one of its lattice vectors, so that the cube becomes a rectangular prism with a square base (a by a) and height (c, which is different from a.) There are two tetragonal Bravais lattices: the simple tetragonal and the centered tetragonal (from stretching either the face-centered or the body-centered cubic lattice.)
Triclinic	In crystallography, the triclinic crystal system is one of the 7 lattice point groups. A crystal system is described by three basis vectors. In the triclinic system, the crystal is described by vectors of unequal length, as in the orthorhombic system.
Crystal system	A Crystal system or crystal family or lattice system is one of six or seven categories of space groups, lattices, point groups, or crystals. Informally, two physical crystals tend to be in the same Crystal system if they have the same symmetries, though there are many exceptions to this. Unfortunately there are several slightly different conventions for the division into seven (or sometimes six) classes, differing mainly in the way crystals with exactly one axis of threefold rotation are classified, and the term Crystal system is used in at least 3 different ways in the literature.
Silver	Silver is a chemical element with the chemical symbol Ag and atomic number 47. A soft, white, lustrous transition metal, it has the highest electrical conductivity of any element and the highest thermal conductivity of any metal. The metal occurs naturally in its pure, free form, as an alloy with gold (electrum) and other metals, and in minerals such as argentite and chlorargyrite.
Potassium	Potassium is a chemical element. It has the symbol K, atomic number 19, and atomic mass 39.0983. Potassium was first isolated from potash.

Chapter 11. States of Matter; Liquids and Solids

Chapter 11. States of Matter; Liquids and Solids

Solid oxide electrolyser cell	A Solid oxide electrolyser cell is a solid oxide fuel cell set in regenerative mode for the electrolysis of water with a solid oxide electrolyte to produce oxygen and hydrogen gas. Solid oxide electrolyser cell s operate at temperatures for high-temperature electrolysis, typically between 500 and 850°C similar to SOFC. Advantages of this class of regenerative fuel cells include high efficiencies, long term stability, fuel flexibility, low emissions, and cost. The largest disadvantage is the high operating temperature which results in longer start up times and mechanical/chemical compatibility issues.
Chromium	Chromium is a chemical element which has the symbol Cr and atomic number 24. It is a steely-gray, lustrous, hard metal that takes a high polish and has a high melting point. It is also odourless, tasteless, and malleable.
Iron	Iron is a chemical element with the symbol Fe and atomic number 26. Iron is a group 8 and period 4 element. Iron and Iron alloys (steels) are by far the most common metals and the most common ferromagnetic materials in everyday use.
Yttrium	Yttrium is a chemical element with symbol Y and atomic number 39. It is a silvery-metallic transition metal chemically similar to the lanthanoids and has historically been classified as a rare earth element. Yttrium is almost always found combined with the lanthanoids in rare earth minerals and is never found in nature as a free element.
Barium	Barium is a chemical element. It has the symbol Ba, and atomic number 56. Barium is a soft silvery metallic alkaline earth metal.
Crystal defects	Crystalline solids have a very regular atomic structure: that is, the local positions of atoms with respect to each other are repeated at the atomic scale. These arrangements are called crystal structures, and their study is called crystallography. However, most crystalline materials are not perfect: the regular pattern of atomic arrangement is interrupted by crystal defects.
Emission	In physics, Emission is the process by which the energy of a photon is released by another entity, for example, by an atom whose electrons make a transition between two electronic energy levels. The emitted energy is in the form of a photon. The emittance of an object quantifies how much light is emitted by it.
Coordination isomerism	Coordination isomerism is a form of structural isomerism in which the composition of the complex ion varies.
Coordination number	The Coordination number of an atom in a molecule or a crystal is the integer number of its nearest neighbours. This number is determined somewhat differently for molecules and for crystals. In chemistry the emphasis is on molecules (or ions) with defined bonding structures, and the Coordination number of an atom is determined by simply counting the other atoms to which it is bonded (by either single or multiple bonds.)

Chapter 11. States of Matter; Liquids and Solids

Chapter 11. States of Matter; Liquids and Solids

Helium	Helium is the chemical element with atomic number 2, and is represented by the symbol He. It is a colorless, odorless, tasteless, non-toxic, inert monatomic gas that heads the noble gas group in the periodic table. Its boiling and melting points are the lowest among the elements and it exists only as a gas except in extreme conditions.
Beryllium	Beryllium is the chemical element with the symbol Be and atomic number 4. A bivalent element, Beryllium is found naturally only combined with other elements in minerals. Notable gemstones which contain Beryllium include Beryl (aquamarines and emeralds) and Chrysoberyl (Alexandrite and Cat"s eye.)
Calcium	Calcium is the chemical element with the symbol Ca and atomic number 20. It has an atomic mass of 40.078 amu. Calcium is a soft grey alkaline earth metal, and is the fifth most abundant element by mass in the Earth"s crust.
Calcium oxide	Calcium oxide, commonly known as burnt lime, lime or quicklime, is a widely used chemical compound. It is a white, caustic and alkaline crystalline solid at room temperature. As a commercial product, lime often also contains magnesium oxide, silicon oxide and smaller amounts of aluminium oxide and iron oxide.
Germanium	Germanium is a chemical element with the symbol Ge and atomic number 32. It is a lustrous, hard, grayish-white metalloid in the carbon group, chemically similar to its group neighbors tin and silicon. Germanium has five naturally occurring isotopes ranging in atomic mass number from 70 to 76.
Thallium	Thallium is a chemical element with the symbol Tl and atomic number 81. This soft gray malleable poor metal resembles tin but discolors when exposed to air. Approximately 60-70% of Thallium production is used in the electronics industry, and the rest is used in the pharmaceutical industry and in glass manufacturing.
Tin	Tin is a chemical element with the symbol Sn and atomic number 50. It is a main group metal in group 14 of the periodic table. Tin shows chemical similarity to both neighboring group 14 elements, germanium and lead, like the two possible oxidation states +2 and +4.
Zinc	Zinc is a metallic chemical element with the symbol Zn and atomic number 30. It is a first-row transition metal in group 12 of the periodic table. Zinc is chemically similar to magnesium because its ion is of similar size and its only common oxidation state is +2.
Zinc oxide	Zinc oxide is an inorganic compound with the formula ZnO. It usually appears as a white powder, nearly insoluble in water. The powder is widely used as an additive into numerous materials and products including plastics, ceramics, glass, cement, rubber (e.g. car tyres), lubricants, paints, ointments, adhesives, sealants, pigments, foods (source of Zn nutrient), batteries, ferrites, fire retardants, etc. ZnO is present in the Earth crust as a mineral zincite; however, most ZnO used commercially is produced synthetically.

Chapter 11. States of Matter; Liquids and Solids

Chapter 11. States of Matter; Liquids and Solids

Gray
: The Gray is the SI unit of absorbed radiation dose due to ionizing radiation (for example, X-rays.) One Gray is the absorption of one joule of energy, in the form of ionizing radiation, by one kilogram of matter.

$$1 \text{ Gy} = 1 \, \frac{\text{J}}{\text{kg}} = 1 \text{ m}^2 \cdot \text{s}^{-2}$$

For X-rays and gamma rays, these are the same units as the sievert (Sv.)

Diffraction
: Diffraction is normally taken to refer to various phenomena which occur when a wave encounters an obstacle. It is described as the apparent bending of waves around small obstacles and the spreading out of waves past small openings. Very similar effects are observed when there is an alteration in the properties of the medium in which the wave is travelling, for example a variation in refractive index for light waves or in acoustic impedance for sound waves and these can also be referred to as Diffraction effects.

Cluster
: In chemistry, a cluster is an ensemble of bound atoms intermediate in size between a molecule and a bulk solid. Clusters exist of diverse stoichiometries and nuclearities. For example, carbon and boron atoms form fullerene and borane clusters, respectively.

Chapter 11. States of Matter; Liquids and Solids

Chapter 12. Solutions

Colloid	A Colloid is a type of chemical mixture where one substance is dispersed evenly throughout another. The particles of the dispersed substance are only suspended in the mixture, unlike a solution, where they are completely dissolved within. This occurs because the particles in a Colloid are larger than in a solution - small enough to be dispersed evenly and maintain a homogenous appearance, but large enough to scatter light and not dissolve.
Solution	In chemistry, a Solution is a homogeneous mixture composed of two or more substances. In such a mixture, a solute is dissolved in another substance, known as a solvent. Gases may dissolve in liquids, for example, carbon dioxide or oxygen in water.
Solvent	A Solvent is a liquid or gas that dissolves a solid, liquid resulting in a solution. The most common Solvent in everyday life is water. Most other commonly-used Solvent s are organic (carbon-containing) chemicals.
Water	Water is the most abundant molecule on Earth"s surface, constituting about 75% of the Earth"s surface. In nature it exists in liquid, solid, and gaseous states. It is in dynamic equilibrium between the liquid and gas states at standard temperature and pressure.
Brine	Brine is water saturated or nearly saturated with a salt (usually sodium chloride.) It is used to preserve vegetables, fish, and meat, in a process known as brining (now less popular than historically.)
Potassium	Potassium is a chemical element. It has the symbol K , atomic number 19, and atomic mass 39.0983. Potassium was first isolated from potash.
Seawater	Seawater is water from a sea or ocean. On average, Seawater in the world"s oceans has a salinity of about 3.5%. This means that every 1 kg of Seawater has approximately 35 grams of dissolved salts (mostly, but not entirely, the ions of sodium chloride: Na^+, Cl^-.)
Sodium	Sodium is a metallic element with a symbol Na and atomic number 11. It is a soft, silvery-white, highly reactive metal and is a member of the alkali metals within "group 1" (formerly known as "group IA".) It has only one stable isotope, ^{23}Na.
Impurities	Impurities are substances inside a confined amount of liquid, gas which differ from the chemical composition of the material or compound. Impurities are either naturally occurring or added during synthesis of a chemical or commercial product. During production, Impurities may be purposely, accidentally, inevitably, or incidentally added into the substance.
Solid solution	A Solid solution is a solid-state solution of one or more solutes in a solvent. Such a mixture is considered a solution rather than a compound when the crystal structure of the solvent remains unchanged by addition of the solutes, and when the mixture remains in a single homogeneous phase.This often happens when the two elements (generally metals) involved are close together on the periodic table; conversely, a chemical compound is generally a result of the non proximity of

the two metals involved on the periodic table.

Chapter 12. Solutions

Chapter 12. Solutions

	The solute may incorporate into the solvent crystal lattice substitutionally, by replacing a solvent particle in the lattice, or interstitial, by fitting into the space between solvent particles.
Acid	An Acid is traditionally considered any chemical compound that, when dissolved in water, gives a solution with a hydrogen ion activity greater than in pure water, i.e. a pH less than 7.0. That approximates the modern definition of Johannes Nicolaus Brønsted and Martin Lowry, who independently defined an Acid as a compound which donates a hydrogen ion (H^+) to another compound (called a base.) Common examples include acetic Acid and sulfuric Acid (used in car batteries.)
Crystallization	Crystallization is the (natural or artificial) process of formation of solid crystals precipitating from a solution, melt or more rarely deposited directly from a gas. Crystallization is also a chemical solid-liquid separation technique, in which mass transfer of a solute from the liquid solution to a pure solid crystalline phase occurs. Frost Crystallization on a shrub. The Crystallization process consists of two major events, nucleation and crystal growth.
Sol	A Sol is a colloidal suspension of solid particles (1 - 500 nanometres in size) in a liquid. Examples include blood, pigmented ink, and paint. Artificial sols may be prepared by dispersion or condensation.
Boiling	Boiling, a type of phase transition, is the rapid vaporization of a liquid, which typically occurs when a liquid is heated to its Boiling point, the temperature at which the vapor pressure of the liquid is equal to the pressure exerted on the liquid by the surrounding environmental pressure. Thus, a liquid may also boil when the pressure of the surrounding atmosphere is sufficiently reduced, such as the use of a vacuum pump or at high altitudes. Boiling occurs in three characteristic stages, which are nucleate, transition and film Boiling.
Melting	Melting is a physical process that results in the phase change of a substance from a solid to a liquid. The internal energy of a solid substance is increased, typically by the application of heat or pressure, resulting in a rise of its temperature to the Melting point, at which the rigid ordering of molecular entities in the solid breaks down to a less-ordered state and the solid liquefies. An object that has melted completely is molten.
Calcium	Calcium is the chemical element with the symbol Ca and atomic number 20. It has an atomic mass of 40.078 amu. Calcium is a soft grey alkaline earth metal, and is the fifth most abundant element by mass in the Earth"s crust.
Barium	Barium is a chemical element. It has the symbol Ba, and atomic number 56. Barium is a soft silvery metallic alkaline earth metal.
Ion	An Ion is an atom or molecule where the total number of electrons is not equal to the total number of protons, giving it a net positive or negative electrical charge.

Chapter 12. Solutions

Chapter 12. Solutions

	Since protons are positively charged and electrons are negatively charged, if there are more electrons than protons, the atom or molecule will be negatively charged. This is called an an Ion, from the Greek á¼€vÎ¬, meaning "up".
Lattice energy	The Lattice energy of an ionic solid is a measure of the strength of bonds in that ionic compound. It is usually defined as the enthalpy of formation of the ionic compound from gaseous ions and as such is invariably exothermic. Na^+ (g) + Cl^- (g) → NaCl (s) The experimental Lattice energy of NaCl is −787 kJ/mol.
Lithium	Lithium is the chemical element with atomic number 3, and is represented by the symbol Li. It is a soft alkali metal with a silver-white color. Under standard conditions it is the lightest metal and the least dense solid element.
Lithium fluoride	Lithium fluoride is a chemical compound of lithium and fluorine. It is a white, inorganic, crystalline, ionic, solid salt under standard conditions. It transmits ultraviolet radiation more efficiently than any other known substance.
Magnesium	Magnesium is a chemical element with the symbol Mg, atomic number 12, atomic weight 24.3050 and common oxidation number +2. Magnesium, an alkaline earth metal, is the ninth most abundant element in the universe by mass. The commonness of Magnesium is related to the fact that it is easily built up in supernova stars from a sequential addition of three helium nuclei to carbon.
Magnesium hydroxide	Magnesium hydroxide is an inorganic compound with the chemical formula $Mg(OH)_2$. As a suspension in water, it may be referred to as Milk of Magnesia. The solid mineral form of Magnesium hydroxide is known as brucite.
Strontium hydroxide	Strontium hydroxide, $Sr(OH)_2$, is a caustic alkali composed of one strontium ion and two hydroxide ions. It is synthesized by combining a strontium salt with a strong base. $Sr(OH)_2$ exists in anhydrous, monohydrate, or octahydrate form.
Hydroxide	In chemistry, Hydroxide is the name for the diatomic anion OH^-, consisting of oxygen and hydrogen atoms, usually derived from the dissociation of a base. It is one of the simplest diatomic ions known. Inorganic compounds that contain the hydroxyl group are referred to as Hydroxide s.
Buffer solution	A Buffer solution is an aqueous solution consisting of a mixture of a weak acid and its conjugate base or a weak base and its conjugate acid. It has the property that the pH of the solution changes very little when a small amount of acid or base is added to it. Buffer solution s are used as a means of keeping pH at a nearly constant value in a wide variety of chemical applications.

Chapter 12. Solutions

Chapter 12. Solutions

Salt	Salt is a dietary mineral composed primarily of sodium chloride that is essential for animal life, but toxic to most land plants. Salt flavor is one of the basic tastes, an important preservative and a popular food seasoning.
Carbon	Carbon is the chemical element with symbol C and atomic number 6. As a member of group 14 on the periodic table, it is nonmetallic and tetravalent--making four electrons available to form covalent chemical bonds. There are three naturally occurring isotopes, with ^{12}C and ^{13}C being stable, while ^{14}C is radioactive, decaying with a half-life of about 5730 years.
Cerium	Cerium is a chemical element with the symbol Ce and atomic number 58. It is a soft, silvery, ductile metal which easily oxidizes in air. Cerium was named after the dwarf planet Ceres.
Copper	Copper is a chemical element with the symbol Cu and atomic number 29. It is a ductile metal with very high thermal and electrical conductivity. Pure Copper is rather soft and malleable and a freshly-exposed surface has a pinkish or peachy color.
Potassium nitrate	Potassium nitrate is a chemical compound with the chemical formula KNO_3. A naturally occurring mineral source of nitrogen, KNO_3 constitutes a critical oxidizing component of black powder/gunpowder. In the past it was also used for several kinds of burning fuses, including slow matches.
Sodium hydroxide	Sodium hydroxide is a caustic metallic base. Sodium hydroxide forms a strong alkaline solution when dissolved in a solvent such as water. However, only the hydroxide ion is basic.
Catalyst	Catalyst is a science centre and museum devoted to the chemical industry. Its full title is Catalyst Science Discovery Centre. It is located in Widnes, Cheshire, in the north west of England, and situated on the north bank of the River Mersey (grid reference SJ512841.)
Electrolysis	In chemistry and manufacturing, Electrolysis is a method of using an electric current to drive an otherwise non-spontaneous chemical reaction. Electrolysis is commercially highly important as a stage in the separation of elements from naturally-occurring sources such as ores using an electrolytic cell. • 1800 - William Nicholson and Johann Ritter decomposed water into hydrogen and oxygen. • 1807 - Potassium, Sodium, Barium, Calcium and Magnesium were discovered by Sir Humphry Davy using Electrolysis. • 1886 - Fluorine was discovered by Henri Moissan using Electrolysis. • 1886 - Hall-Héroult process developed for making aluminium • 1890 - Castner-Kellner process developed for making sodium hydroxide

Chapter 12. Solutions

Chapter 12. Solutions

Electrolysis is the passage of an electric current through an ionic substance that is either molten or dissolved in a suitable solvent, resulting in chemical reactions at the electrodes and separation of materials.

The main components required to achieve Electrolysis are:

- A liquid containing mobile ions - an electrolyte
- An external source of direct electric current
- Two solid rods or plates known as electrodes

The components perform the following roles in the Electrolysis process:

- The mobile ions are the carriers of electrical current in the liquid (electrolyte.) If the ions are not mobile, as in a solid salt then Electrolysis cannot occur.

Principle	A Principle is a comprehensive and fundamental law, doctrine, or assumption. It can be a rule or code of conduct. It can be a law or fact of nature underlying the working of an artificial device.
Colligative properties	Colligative properties are properties of solutions that depend on the number of molecules in a given volume of solvent and not on the properties (e.g. size or mass) of the molecules. Colligative properties include: lowering of vapor pressure; elevation of boiling point; depression of freezing point and osmotic pressure. Measurements of these properties for a dilute aqueous solution of a non-ionized solute such as urea or glucose can lead to accurate determinations of relative molecular masses.
Concentration	In chemistry, Concentration is the measure of how much of a given substance there is mixed with another substance. This can apply to any sort of chemical mixture, but most frequently the concept is limited to homogeneous solutions, where it refers to the amount of solute in the solvent. To concentrate a solution, one must add more solute, or reduce the amount of solvent (for instance, by selective evaporation.)
Freezing-point depression	Freezing-point depression describes the phenomenon that the freezing point of a liquid (a solvent) is depressed when another compound is added, meaning that a solution has a lower freezing point than a pure solvent. This happens whenever a solute is added to a pure solvent, such as water. The phenomenon may be observed in sea water, which due to its salt content remains liquid at temperatures below 0°C, the freezing point of pure water.
Molar concentration	In chemistry, molar concentration is a measure of the concentration of a solute in a solution ionic in thermodynamics the use of molar concentration is often not very convenient, because the volume of most solutions slightly depends on temperature due to thermal expansion. This problem is usually resolved by introducing temperature correction factors, or by using a temperature-independent measure of concentration such as molality.

Chapter 12. Solutions

Chapter 12. Solutions

Aqueous solution	An Aqueous solution is a solution in which the solvent is water. It is usually shown in chemical equations by appending (aq) to the relevant formula. The word aqueous means pertaining to, related to, similar to, or dissolved in water.
Freezing	In physical science, Freezing or solidification is the process in which a liquid turns into a solid when cold enough. The Freezing point is the temperature at which this happens. Melting, the process of turning a solid to a liquid, is almost the exact opposite of Freezing.
Mole	The mole is a unit of amount of substance: it is an SI base unit, and one of the few units used to measure this physical quantity. The name "mole" was coined in German by Wilhelm Ostwald in 1893, although the related concept of equivalent mass had been in use at least a century earlier. The name is assumed to be derived from the word Molekül (molecule.)
Mole fraction	In chemistry, mole fraction x (also, and more correctly, known as the amount fraction) is a way of expressing the composition of a mixture. The mole fraction of each component i is defined as its amount of substance n_i divided by the total amount of substance in the system, n $$x_i \stackrel{def}{=} \frac{n_i}{n}$$ where $$n = \sum_i n_i$$ The sum is over all components, including the solvent in the case of a chemical solution. As an example, if a mixture is obtained by dissolving 10 moles of sucrose in 90 moles of water, the mole fraction of sucrose in that mixture is 0.1.
Fraction	A Fraction in chemistry is a quantity collected from a sample or batch of a substance in a fractionating separation process. In such a process, a mixture is separated into fractions, which have compositions that vary according to a gradient. A Fraction can be defined as a group of chemicals that have similar boiling points.
Fractional distillation	Fractional distillation is the separation of a mixture into its component parts such as in separating chemical compounds by their boiling point by heating them to a temperature at which several fractions of the compound will evaporate. It is a special type of distillation. Generally the component parts boil at less than 25 °C from each other under a pressure of one atmosphere (atm.)
Lead	Lead is a main-group element with symbol Pb and atomic number 82. Lead is a soft, malleable poor metal, also considered to be one of the heavy metals. Lead has a bluish-white color when freshly cut, but tarnishes to a dull grayish color when exposed to air.

Chapter 12. Solutions

Chapter 12. Solutions

Vapor	A Vapor or vapour is a substance in the gas phase at a temperature lower than its critical temperature. This means that the Vapor can be condensed to a liquid or to a solid by increasing its pressure, without reducing the temperature. For example, water has a critical temperature of 374°C (or 647 K) which is the highest temperature at which liquid water can exist.
Vapor pressure	Vapor pressure, is the pressure of a vapor in equilibrium with its non-vapor phases. All liquids and solids have a tendency to evaporate to a gaseous form, and all gases have a tendency to condense back into their original form At any given temperature, for a particular substance, there is a pressure at which the gas of that substance is in dynamic equilibrium with its liquid or solid forms.
Benzene	Benzene, or benzol, is an organic chemical compound and a known carcinogen with the molecular formula C_6H_6. It is sometimes abbreviated Ph-H. Benzene is a colorless and highly flammable liquid with a sweet smell and a relatively high melting point. Because it is a known carcinogen, its use as an additive in gasoline is now limited, but it is an important industrial solvent and precursor in the production of drugs, plastics, synthetic rubber, and dyes.
Toluene	Toluene phenylmethane, and Toluol, is a clear water-insoluble liquid with the typical smell of paint thinners, redolent of the sweet smell of the related compound benzene. It is an aromatic hydrocarbon that is widely used as an industrial feedstock and as a solvent. Like other solvents, Toluene is also used as an inhalant drug for its intoxicating properties; however this causes severe neurological harm.
Chlorination	Chlorination is the process of adding the element chlorine to water as a method of water purification to make it fit for human consumption as drinking water. Water which has been treated with chlorine is effective in preventing the spread of disease. The Chlorination of public drinking supplies was originally met with resistance, as people were concerned about the health effects of the practice.
Distillation	Distillation is a method of separating mixtures based on differences in their volatilities in a boiling liquid mixture. Distillation is a unit operation, or a physical separation process, and not a chemical reaction. Commercially, Distillation has a number of uses.
Boiling-point elevation	Boiling-point elevation describes the phenomenon that the boiling point of a liquid (a solvent) will be higher when another compound is added, meaning that a solution has a higher boiling point than a pure solvent. This happens whenever a non-volatile solute, such as a salt, is added to a pure solvent, such as water. The boiling point can be measured accurately using an ebullioscope.

Chapter 12. Solutions

Chapter 12. Solutions

Molecular mass	The Molecular mass of a substance, frequently referred by the older term molecular weight and abbreviated as MW, is the mass of one molecule of that substance, relative to the unified atomic mass unit u (equal to 1/12 the mass of one isotope of carbon-12.) This is distinct from the relative Molecular mass of a molecule, which is the ratio of the mass of that molecule to 1/12 of the mass of carbon 12 and is a dimensionless number. Relative Molecular mass is abbreviated to M_r.
Melvin Ellis Calvin	Melvin Ellis Calvin was an American chemist most famed for discovering the Calvin cycle along with Andrew Benson and James Bassham, for which he was awarded the 1961 Nobel Prize in Chemistry. He spent most of his five-decade career at the University of California, Berkeley. Calvin was born in St. Paul, Minnesota, the son of Jewish immigrants.
Osmotic pressure	Osmotic pressure is the hydrostatic pressure produced by a difference in concentration between solutions on the two sides of a surface such as a semipermeable membrane. Jacobus Henricus van "t Hoff first proposed a formula for calculating the Osmotic pressure, but this was later improved upon by Harmon Northrop Morse. A related notion, osmotic potential is the opposite of water potential, with the former meaning the degree to which a solvent would want to stay in a liquid.
Polyethylene	Polyethylene or polythene (IUPAC name polyethene or poly(methylene)) is a thermoplastic commodity heavily used in consumer products (notably the plastic shopping bag.) Over 60 million tons of the material are produced worldwide every year. Polyethylene is a polymer consisting of long chains of the monomer ethylene (IUPAC name ethene.)
Polymer	A Polymer is a large molecule composed of repeating structural units typically connected by covalent chemical bonds. While Polymer in popular usage suggests plastic, the term actually refers to a large class of natural and synthetic materials with a variety of properties. Due to the extraordinary range of properties accessible in Polymer ic materials , they have come to play an essential and ubiquitous role in everyday life - from plastics and elastomers on the one hand to natural bio Polymer s such as DNA and proteins that are essential for life on the other.
Emulsion	An Emulsion (IPA: /ÉªËˆmÊŒlÊƒÉ™n/) is a mixture of two or more immiscible (unblendable) liquids. One liquid (the dispersed phase) is dispersed in the other (the continuous phase.) Many Emulsion s are oil/water Emulsion s, with dietary fats being one common type of oil encountered in everyday life.
Tyndall effect	The Tyndall effect is an effect of light scattering by colloidal particles or particles in suspension. It is named after the 19th century Irish scientist John Tyndall. It is similar to Rayleigh scattering, in that the intensity of the scattered light depends on the fourth power of the frequency, so blue light is scattered more strongly than red light.

Chapter 12. Solutions

Chapter 12. Solutions

Iron	Iron is a chemical element with the symbol Fe and atomic number 26. Iron is a group 8 and period 4 element. Iron and Iron alloys (steels) are by far the most common metals and the most common ferromagnetic materials in everyday use.
Micelle	A Micelle , plural Micelle s, micella, or micellae) is an aggregate of surfactant molecules dispersed in a liquid colloid. A typical Micelle in aqueous solution forms an aggregate with the hydrophilic "head" regions in contact with surrounding solvent, sequestering the hydrophobic single tail regions in the Micelle centre. This phase is caused by the insufficient packing issues of single tailed lipids in a bilayer.
Soap	Soap is an anionic surfactant used in conjunction with water for washing and cleaning, which historically comes either in solid bars or in the form of a viscous liquid. Soap consists of sodium or potassium salts of fatty acids and is obtained by reacting common oils or fats with a strong alkaline solution (the base, popularly referred to as lye) in a process known as saponification. The fats are hydrolyzed by the base, yielding alkali salts of fatty acids (crude Soap) and glycerol.
Detergent	A Detergent is a material intended to assist cleaning. The term is sometimes used to differentiate between soap and other surfactants used for cleaning. As an adjective pertaining to a substance, it (or "detersive") means "cleaning" or "having cleaning properties"; "detergency" indicates presence or degree of cleaning property.
Linkage isomerism	Linkage isomerism is the existence of co-ordination compounds that have the same composition differing with the connectivity of the metal to a ligand. The first reported example of Linkage isomerism had the formula $[Co(NH_3)_5(NO_2)]Cl_2$. The cationic cobalt complex exists in two separable linkage isomers.

Chapter 12. Solutions

Chapter 13. Rates of Reaction

Barium	Barium is a chemical element. It has the symbol Ba, and atomic number 56. Barium is a soft silvery metallic alkaline earth metal.
Concentration	In chemistry, Concentration is the measure of how much of a given substance there is mixed with another substance. This can apply to any sort of chemical mixture, but most frequently the concept is limited to homogeneous solutions, where it refers to the amount of solute in the solvent. To concentrate a solution, one must add more solute, or reduce the amount of solvent (for instance, by selective evaporation.)
Empirical formula	In chemistry, the Empirical formula of a chemical compound is a simple expression of the relative numbers of each type of atom in it, or the simplest whole number ratio of atoms of each element present in a compound. An Empirical formula makes no reference to isomerism, structure, or absolute number of atoms. The Empirical formula is used as standard for most ionic compounds, such as $CaCl_2$, and for macromolecules, such as SiO_2.
Nitrogen	Nitrogen is a chemical element that has the symbol N and atomic number 7 and atomic mass 14.00674 u. Elemental Nitrogen is a colorless, odorless, tasteless and mostly inert diatomic gas at standard conditions, constituting 78% by volume of Earth"s atmosphere. Many industrially important compounds, such as ammonia, nitric acid, organic nitrates , and cyanides, contain Nitrogen.
Yield	Nuclear fission splits a heavy nucleus such as uranium or plutonium into two lighter nuclei, which are called fission products. Yield refers to the fraction of a fission product produced per fission. Yield can be broken down by: 1. Individual isotope 2. Chemical element spanning several isotopes of different mass number but same atomic number. 3. Nuclei of a given mass number regardless of atomic number. Known as "chain Yield" because it represents a decay chain of beta decay. Isotope and element yields will change as the fission products undergo beta decay, while chain yields do not change after completion of neutron emission by a few neutron-rich initial fission products (delayed neutrons), with halflife measured in seconds. A few isotopes can be produced directly by fission, but not by beta decay because the would-be precursor with atomic number one greater is stable and does not decay.
Catalyst	Catalyst is a science centre and museum devoted to the chemical industry. Its full title is Catalyst Science Discovery Centre. It is located in Widnes, Cheshire, in the north west of England, and situated on the north bank of the River Mersey (grid reference SJ512841.)

Chapter 13. Rates of Reaction

Chapter 13. Rates of Reaction

Hydrobromic acid	Hydrobromic acid is a strong acid formed by dissolving the diatomic molecule hydrogen bromide in water. "Constant boiling" Hydrobromic acid is an aqueous solution that distills at 124.3 °C and contains 47.6% HBr by weight. Hydrobromic acid has a pK_a of −9, making it a stronger acid than hydrochloric acid, but not as strong as hydroiodic acid.
Hydrogen	Hydrogen is the chemical element with atomic number 1. It is represented by the symbol H. At standard temperature and pressure, Hydrogen is a colorless, odorless, nonmetallic, tasteless, highly flammable diatomic gas with the molecular formula H_2. With an atomic weight of 1.007 94 u, Hydrogen is the lightest element.
Acid	An Acid is traditionally considered any chemical compound that, when dissolved in water, gives a solution with a hydrogen ion activity greater than in pure water, i.e. a pH less than 7.0. That approximates the modern definition of Johannes Nicolaus Brønsted and Martin Lowry, who independently defined an Acid as a compound which donates a hydrogen ion (H^+) to another compound (called a base.) Common examples include acetic Acid and sulfuric Acid (used in car batteries.)
Impurities	Impurities are substances inside a confined amount of liquid, gas which differ from the chemical composition of the material or compound.
	Impurities are either naturally occurring or added during synthesis of a chemical or commercial product. During production, Impurities may be purposely, accidentally, inevitably, or incidentally added into the substance.
Fluorine	Fluorine is a chemical element, represented by the symbol F, and the atomic number 9. Fluorine forms a single bond with itself in elemental form, resulting in the diatomic F_2 molecule. F_2 is a supremely reactive, poisonous, pale, yellowish brown gas.
Infrared spectroscopy	Infrared spectroscopy is the subset of spectroscopy that deals with the infrared region of the electromagnetic spectrum. It covers a range of techniques, the most common being a form of absorption spectroscopy. As with all spectroscopic techniques, it can be used to identify compounds or investigate sample composition.
Nuclear magnetic resonance	Nuclear magnetic resonance is a property that magnetic nuclei have in a magnetic field and applied electromagnetic (EM) pulse, which cause the nuclei to absorb energy from the EM pulse and radiate this energy back out. The energy radiated back out is at a specific resonance frequency which depends on the strength of the magnetic field and other factors. This allows the observation of specific quantum mechanical magnetic properties of an atomic nucleus.

Chapter 13. Rates of Reaction

Chapter 13. Rates of Reaction

Resonance	Resonance in chemistry is a key component of valence bond theory used to graphically represent and mathematically model certain types of molecular structures when no single, conventional Lewis structure can satisfactorily represent the observed structure or explain its properties. Resonance instead considers such molecules to be an intermediate or average (called a Resonance hybrid) between several Lewis structures that differ only in the placement of the valence electrons. Scheme 1.
Spectroscopy	Spectroscopy was originally the study of the interaction between radiation and matter as a function of wavelength (λ.) In fact, historically, Spectroscopy referred to the use of visible light dispersed according to its wavelength, e.g. by a prism. Later the concept was expanded greatly to comprise any measurement of a quantity as function of either wavelength or frequency.
Boiling	Boiling, a type of phase transition, is the rapid vaporization of a liquid, which typically occurs when a liquid is heated to its Boiling point, the temperature at which the vapor pressure of the liquid is equal to the pressure exerted on the liquid by the surrounding environmental pressure. Thus, a liquid may also boil when the pressure of the surrounding atmosphere is sufficiently reduced, such as the use of a vacuum pump or at high altitudes. Boiling occurs in three characteristic stages, which are nucleate, transition and film Boiling.
Order	Order in a crystal lattice is the arrangement of some property with respect to atomic positions. It arises in charge ordering, spin ordering, magnetic ordering, and compositional ordering. It is a thermodynamic entropy concept often displayed by a second Order phase transition.
Enthalpy	In thermodynamics and molecular chemistry, the Enthalpy is a thermodynamic property of a fluid. It can be used to calculate the heat transfer during a quasistatic process taking place in a closed thermodynamic system under constant pressure. Enthalpy H is an arbitrary concept but the Enthalpy change ΔH is more useful because it is equal to the change in the internal energy of the system, plus the work that the system has done on its surroundings.
Iodine	Iodine , is a chemical element that has the symbol I and atomic number 53. Naturally-occurring Iodine is a single isotope with 74 neutrons. Chemically, Iodine is the second least reactive of the halogens, and the second most electropositive halogen; trailing behind astatine in both of these categories.
Ion	An Ion is an atom or molecule where the total number of electrons is not equal to the total number of protons, giving it a net positive or negative electrical charge. Since protons are positively charged and electrons are negatively charged, if there are more electrons than protons, the atom or molecule will be negatively charged. This is called an an Ion , from the Greek á¼€νÎ¬ , meaning "up".

Chapter 13. Rates of Reaction

Chapter 13. Rates of Reaction

Oxoacid	An Oxoacid is an acid which contains oxygen. More specifically, it is an acid which: 1. contains oxygen; 2. contains at least one other element; 3. has at least one hydrogen atom bound to oxygen; and 4. forms an ion by the loss of one or more protons. The name oxyacid is sometimes used, although this is not recommended. Generally, Oxoacid s are simply polyatomic ions with a hydrogen cation. Under Lavoisier"s original theory, all acids contained oxygen, which was named from the Greek οξυς (acid, sharp) and γεινομαι (geinomai) (engender.)
Thymol blue	Thymol blue is a brownish-green or reddish-brown crystaline powder that is used as an pH indicator. It is insoluble in water but soluble in alcohol and dilute alkali solutions. It transitions from red to yellow at pH 1.2-2.8 and from yellow to blue from at pH 8.0-9.6.
Critical opalescence	Critical opalescence is a phenomenon which arises in the region of a continuous phase transition. Originally reported by Thomas Andrews in 1869 for the liquid-gas transition in carbon dioxide, many other examples have been discovered since. The phenomenon is most commonly demonstrated in binary fluid mixtures, such as methanol and cyclohexane.
Half-life	The Half-life of a quantity whose value decreases with time is the interval required for the quantity to decay to half of its initial value. The concept originated in describing how long it takes atoms to undergo radioactive decay but also applies in a wide variety of other situations. The term "Half-life" dates to 1907.
Radioactive decay	Radioactive decay is the process in which an unstable atomic nucleus spontaneously loses energy by emitting ionizing particles and radiation. This decay, or loss of energy, results in an atom of one type, called the parent nuclide transforming to an atom of a different type, called the daughter nuclide. For example: a carbon-14 atom (the "parent") emits radiation and transforms to a nitrogen-14 atom (the "daughter".)
Chlorin	In organic chemistry, a Chlorin is a large heterocyclic aromatic ring consisting, at the core, of three pyrroles and one pyrroline coupled through four methine linkages. Unlike a porphyrin, a Chlorin is therefore largely aromatic but not aromatic through the entire circumference of the ring. Magnesium-containing Chlorin s are called chlorophylls, and are the central photosensitive pigment in chloroplasts.
Chlorine	Chlorine . As the chloride ion, which is part of common salt and other compounds, it is abundant in nature and necessary to most forms of life, including humans. In its elemental form (Cl_2 or "di Chlorine ") under standard conditions, Chlorine is a powerful oxidant and is used in bleaching and disinfectants.

Chapter 13. Rates of Reaction

Chapter 13. Rates of Reaction

Complex	In chemistry, a Complex is a structure consisting of a central atom or molecule connected to surrounding atoms or molecules. Originally, a Complex implied a reversible association of molecules, atoms, or ions through weak chemical bonds. As applied to coordination chemistry, this meaning has evolved.
Activation energy	In chemistry, Activation energy is a term introduced in 1889 by the Swedish scientist Svante Arrhenius, that is defined as the energy that must be overcome in order for a chemical reaction to occur. Arrhenius" research was a follow up of the theories of reaction rate by Serbian physicist Nebojsa Lekovic. Activation energy may also be defined as the minimum energy required to start a chemical reaction.
Hydrogen iodide	Hydrogen iodide is a diatomic molecule. Aqueous solutions of Hydrogen iodide are known as iohydroic acid or hydriodic acid, a strong acid. Hydrogen iodide and hydroiodic acid are, however, different in that the former is a gas under standard conditions; whereas, the other is an aqueous solution of said gas.
Carbon	Carbon is the chemical element with symbol C and atomic number 6. As a member of group 14 on the periodic table, it is nonmetallic and tetravalent--making four electrons available to form covalent chemical bonds. There are three naturally occurring isotopes, with ^{12}C and ^{13}C being stable, while ^{14}C is radioactive, decaying with a half-life of about 5730 years.
Reaction mechanism	In chemistry, a Reaction mechanism is the step by step sequence of elementary reactions by which overall chemical change occurs . Although only the net chemical change is directly observable for most chemical reactions, experiments can often be designed that suggest the possible sequence of steps in a Reaction mechanism. A mechanism describes in detail exactly what takes place at each stage of a chemical transformation.
Acid-base reaction	An Acid-base reaction is a chemical reaction that occurs between an acid and a base. Several concepts that provide alternative definitions for the reaction mechanisms involved and their application in solving related problems exist. Despite several differences in definitions, their importance becomes apparent as different methods of analysis when applied to Acid-base reaction s for gaseous or liquid species, or when acid or base character may be somewhat less apparent.
Reaction intermediate	A Reaction intermediate or an intermediate is a molecular entity that is formed from the reactants (or preceding intermediates) and reacts further to give the directly observed products of a chemical reaction. Most chemical reactions are stepwise, that is they take more than one elementary step to complete. An intermediate is the reaction product of each of these steps, except for the last one, which forms the final product.

Chapter 13. Rates of Reaction

Chapter 13. Rates of Reaction

Chlorination	Chlorination is the process of adding the element chlorine to water as a method of water purification to make it fit for human consumption as drinking water. Water which has been treated with chlorine is effective in preventing the spread of disease. The Chlorination of public drinking supplies was originally met with resistance, as people were concerned about the health effects of the practice.
Bromine	Bromine , Greek: βρά¿¶μος, brómos, meaning "stench "), is a chemical element with the symbol Br and atomic number 35. A halogen element, Bromine is a reddish-brown volatile liquid at standard room temperature that is intermediate in reactivity between chlorine and iodine. Bromine vapours are corrosive and toxic.
Ozone	Ozone or trioxygen (O_3) is a triatomic molecule, consisting of three oxygen atoms. It is an allotrope of oxygen that is much less stable than the diatomic O_2. Ground-level Ozone is an air pollutant with harmful effects on the respiratory systems of animals.
Molecule	A Molecule is defined as a sufficiently stable, electrically neutral group of at least two atoms in a definite arrangement held together by very strong (covalent) chemical bonds. Molecule s are distinguished from polyatomic ions in this strict sense. In organic chemistry and biochemistry, the term Molecule is used less strictly and also is applied to charged organic Molecule s and bio Molecule s.
Catalysis	Catalysis is the process in which the rate of a chemical reaction is either increased or decreased by means of a chemical substance known as a catalyst. Unlike other reagents that participate in the chemical reaction, a catalyst is not consumed by the reaction itself. The catalyst may participate in multiple chemical transformations.
Enzymes	Enzymes are biomolecules that catalyze (i.e., increase the rates of) chemical reactions. Nearly all known Enzymes are proteins. However, certain RNA molecules can be effective biocatalysts too.
Enzyme catalysis	Enzyme catalysis is the catalysis of chemical reactions by specialized proteins known as enzymes. Catalysis of biochemical reactions in the cell is vital due to the very low reaction rates of the uncatalysed reactions. The mechanism of Enzyme catalysis is similar in principle to other types of chemical catalysis.
Cerium	Cerium is a chemical element with the symbol Ce and atomic number 58. It is a soft, silvery, ductile metal which easily oxidizes in air. Cerium was named after the dwarf planet Ceres.
Homogeneous catalysis	Homogeneous catalysis is a chemistry term which describes catalysis where the catalyst is in the same state (ie. solid, liquid and gas), as the reactants. It is the opposite to heterogeneous catalysis.

Chapter 13. Rates of Reaction

Chapter 13. Rates of Reaction

Sulfur	Sulfur or sulphur is the chemical element that has the atomic number 16. It is denoted with the symbol S. It is an abundant, multivalent non-metal. Sulfur in its native form is a yellow crystalline solid.
Thallium	Thallium is a chemical element with the symbol Tl and atomic number 81. This soft gray malleable poor metal resembles tin but discolors when exposed to air. Approximately 60-70% of Thallium production is used in the electronics industry, and the rest is used in the pharmaceutical industry and in glass manufacturing.
Acid rain	Acid rain is rain or any other form of precipitation that is unusually acidic. It has harmful effects on plants, aquatic animals, and infrastructure. Acid rain is mostly caused by human emissions of sulfur and nitrogen compounds which react in the atmosphere to produce acids.
Adsorption	Adsorption is the accumulation of atoms or molecules on the surface of a material. This process creates a film of the adsorbate (the molecules or atoms being accumulated) on the adsorbent"s surface. It is different from absorption, in which a substance diffuses into a liquid or solid to form a solution.
Chemisorption	Chemisorption is a classification of adsorption characterized by a strong interaction between an adsorbate and a substrate surface, as opposed to physisorption which is characterized by a weak Van der Waals force. A distinction between the two can be difficult and it is conventionally accepted that it is around 0.5 eV of binding energy per atom or molecule. The types of strong interactions include chemical bonds of the ionic or covalent variety, depending on the species involved.
Contact process	The Contact process is the current method of producing sulfuric acid in the high concentrations needed for industrial processes. Vanadium(V) oxide (vanadium pentoxide) is the catalyst employed. This process was patented in 1831 by the British vinegar merchant Peregrine Phillips.
Heterogeneous catalysis	Heterogeneous catalysis is a chemistry term which describes catalysis where the catalyst is in a different phase (ie. solid, liquid and gas, but also oil and water) to the reactants. Heterogeneous catalysts provide a surface on which the reaction may take place.
Hydrogenation	Hydrogenation is the chemical reaction that results from the addition of hydrogen (H_2.) The process is usually employed to reduce or saturate organic compounds. The process typically constitutes the addition of pairs of hydrogen atoms to a molecule.

Chapter 13. Rates of Reaction

Catalytic converter	A Catalytic converter (colloquially, "cat" or "catcon") is a device used to reduce the toxicity of emissions from an internal combustion engine. First widely introduced on series-production automobiles in the U.S. market for the 1975 model year to comply with tightening EPA regulations on auto exhaust, Catalytic converter s are still most commonly used in motor vehicle exhaust systems. Catalytic converter s are also used on generator sets, forklifts, mining equipment, trucks, buses, trains, and other engine-equipped machines.
Polymer	A Polymer is a large molecule composed of repeating structural units typically connected by covalent chemical bonds. While Polymer in popular usage suggests plastic, the term actually refers to a large class of natural and synthetic materials with a variety of properties. Due to the extraordinary range of properties accessible in Polymer ic materials , they have come to play an essential and ubiquitous role in everyday life - from plastics and elastomers on the one hand to natural bio Polymer s such as DNA and proteins that are essential for life on the other.
Active site	The Active site of an enzyme contains the catalytic and binding sites. The structure and chemical properties of the Active site allow the recognition and binding of the substrate. The Active site is usually a big pocket or cleft surrounded by amino acid- and other side chains at the surface of the enzyme that contains residues responsible for the substrate specificity (charge, hydrophobicity, steric hindrance) and catalytic residues which often act as proton donors or acceptors or are responsible for binding a cofactor such as PLP, TPP or NAD. The Active site is also the site of inhibition of enzymes
Substrate	Substrate is a term used in materials science to describe the base material on which processing is conducted to produce new film or layers of material such as deposited coatings.

Chapter 13. Rates of Reaction

Chapter 13. Rates of Reaction

A typical Substrate might be a metal, onto which a coating might be deposited by any of the following processes:

- Coating and printing processes
- Chemical vapor deposition and physical vapor deposition
- Conversion coating
    - Anodizing
    - Chromate conversion coating
    - Plasma electrolytic oxidation
    - Phosphate (coating)
- Paint
    - Enamel (paint)
    - Powder coating
    - Industrial coating
    - Silicate mineral paint
    - Fusion bonded epoxy coating (FBE coating)
- Pickled and oiled, a type of plate steel coating.
- Plasma coatings
- Plating
    - Electroless plating
    - Electrochemical plating
- Polymer coatings, such as Teflon
- Sputtered or vacuum deposited materials
- Enamel

In optics, glass may be used as a Substrate for an optical coating--either an antireflection coating to reduce reflection, or a mirror coating to enhance it.

A Substrate may be also an engineered surface where an unintended or natural process occurs, like in:

- Fouling
- Corrosion
- Biofouling
- Heterogeneous catalysis
- Adsorption

Chapter 13. Rates of Reaction

Chapter 13. Rates of Reaction

Sucrose	Sucrose is a disaccharide of glucose and fructose with an α 1,2 glycosidic linkage. The molecular formula of Sucrose is $C_{12}H_{22}O_{11}$. Its systematic name is β-D-fructofuranosyl--α-D-glucopyranoside
Crystal	A Crystal or Crystal line solid is a solid material whose constituent atoms, molecules, or ions are arranged in an orderly repeating pattern extending in all three spatial dimensions. The scientific study of Crystal s and Crystal formation is Crystal lography. The process of Crystal formation via mechanisms of Crystal growth is called Crystal lization or solidification.
Crystal structure	In mineralogy and crystallography, a Crystal structure is a unique arrangement of atoms in a crystal. A Crystal structure is composed of a motif, a set of atoms arranged in a particular way, and a lattice. Motifs are located upon the points of a lattice, which is an array of points repeating periodically in three dimensions.

Chapter 13. Rates of Reaction

Chapter 14. Chemical Equilibrium

Carbon	Carbon is the chemical element with symbol C and atomic number 6. As a member of group 14 on the periodic table, it is nonmetallic and tetravalent--making four electrons available to form covalent chemical bonds. There are three naturally occurring isotopes, with ^{12}C and ^{13}C being stable, while ^{14}C is radioactive, decaying with a half-life of about 5730 years.
Hydrogen	Hydrogen is the chemical element with atomic number 1. It is represented by the symbol H. At standard temperature and pressure, Hydrogen is a colorless, odorless, nonmetallic, tasteless, highly flammable diatomic gas with the molecular formula H_2. With an atomic weight of 1.007 94 u, Hydrogen is the lightest element.
Acid-base reaction	An Acid-base reaction is a chemical reaction that occurs between an acid and a base. Several concepts that provide alternative definitions for the reaction mechanisms involved and their application in solving related problems exist. Despite several differences in definitions, their importance becomes apparent as different methods of analysis when applied to Acid-base reaction s for gaseous or liquid species, or when acid or base character may be somewhat less apparent.
Methanation	Methanation is a physical-chemical process to generate Methane from a mixture of various gases out of biomass fermentation or thermo-chemical gasification. The main components are carbon monoxide and hydrogen. The following main process describes the Methanation: $$CO + 3\,H_2 \rightarrow CH_4 + H_2O$$ This process is used for the generation of biogenous natural gas substitute, which can be fed into the gas grid.Methanation is the reverse reaction of steam methane reforming, which converts methane into synthesis gas.
Stoichiometry	Stoichiometry is the calculation of quantitative (measurable) relationships of the reactants and products in a balanced chemical reaction (chemicals.) It can be used to calculate quantities such as the amount of products that can be produced with the given reactants and percent yield. "Stoichiometry" is derived from the Greek words στοιχεά¿–ον and μι̂τρον (metron, meaning measure.)
Atomic mass	The Atomic mass is the mass of an atom, most often expressed in unified Atomic mass units. The Atomic mass may be considered to be the total mass of protons, neutrons and electrons in a single atom (when the atom is motionless.) The Atomic mass is sometimes incorrectly used as a synonym of relative Atomic mass, average Atomic mass and atomic weight; however, these differ subtly from the Atomic mass.
Law of mass action	In chemistry, the Law of mass action explains behaviors of solutions in dynamic equilibrium. It can be described with two aspects: 1) the equilibrium aspect, concerning the composition of a reaction mixture at equilibrium and 2) the kinetic aspect concerning the rate equations for elementary reactions. Both aspects stem from the research by Guldberg and Waage (1864-1879) in which equilibrium constants were derived by using kinetic data and the rate equation which they had proposed.

Chapter 14. Chemical Equilibrium

Chapter 14. Chemical Equilibrium

Molecular mass	The Molecular mass of a substance, frequently referred by the older term molecular weight and abbreviated as MW, is the mass of one molecule of that substance, relative to the unified atomic mass unit u (equal to 1/12 the mass of one isotope of carbon-12.) This is distinct from the relative Molecular mass of a molecule, which is the ratio of the mass of that molecule to 1/12 of the mass of carbon 12 and is a dimensionless number. Relative Molecular mass is abbreviated to M_r.
Nitrogen	Nitrogen is a chemical element that has the symbol N and atomic number 7 and atomic mass 14.00674 u. Elemental Nitrogen is a colorless, odorless, tasteless and mostly inert diatomic gas at standard conditions, constituting 78% by volume of Earth"s atmosphere. Many industrially important compounds, such as ammonia, nitric acid, organic nitrates , and cyanides, contain Nitrogen.
Concentration	In chemistry, Concentration is the measure of how much of a given substance there is mixed with another substance. This can apply to any sort of chemical mixture, but most frequently the concept is limited to homogeneous solutions, where it refers to the amount of solute in the solvent. To concentrate a solution, one must add more solute, or reduce the amount of solvent (for instance, by selective evaporation.)
Hydrogen iodide	Hydrogen iodide is a diatomic molecule. Aqueous solutions of Hydrogen iodide are known as iohydroic acid or hydriodic acid, a strong acid. Hydrogen iodide and hydroiodic acid are, however, different in that the former is a gas under standard conditions; whereas, the other is an aqueous solution of said gas.
Iron	Iron is a chemical element with the symbol Fe and atomic number 26. Iron is a group 8 and period 4 element. Iron and Iron alloys (steels) are by far the most common metals and the most common ferromagnetic materials in everyday use.
Calcium	Calcium is the chemical element with the symbol Ca and atomic number 20. It has an atomic mass of 40.078 amu. Calcium is a soft grey alkaline earth metal, and is the fifth most abundant element by mass in the Earth"s crust.
Calcium oxide	Calcium oxide, commonly known as burnt lime, lime or quicklime, is a widely used chemical compound. It is a white, caustic and alkaline crystalline solid at room temperature. As a commercial product, lime often also contains magnesium oxide, silicon oxide and smaller amounts of aluminium oxide and iron oxide.
Ammonia	Ammonia is a compound of nitrogen and hydrogen with the formula NH_3. It is normally encountered as a gas with a characteristic pungent odor. Ammonia contributes significantly to the nutritional needs of terrestrial organisms by serving as a precursor to foodstuffs and fertilizers.

Chapter 14. Chemical Equilibrium

Chapter 14. Chemical Equilibrium

Aqueous solution	An Aqueous solution is a solution in which the solvent is water. It is usually shown in chemical equations by appending (aq) to the relevant formula. The word aqueous means pertaining to, related to, similar to, or dissolved in water.
Solution	In chemistry, a Solution is a homogeneous mixture composed of two or more substances. In such a mixture, a solute is dissolved in another substance, known as a solvent. Gases may dissolve in liquids, for example, carbon dioxide or oxygen in water.
Water	Water is the most abundant molecule on Earth"s surface, constituting about 75% of the Earth"s surface. In nature it exists in liquid, solid, and gaseous states. It is in dynamic equilibrium between the liquid and gas states at standard temperature and pressure.
Iodine	Iodine , is a chemical element that has the symbol I and atomic number 53. Naturally-occurring Iodine is a single isotope with 74 neutrons. Chemically, Iodine is the second least reactive of the halogens, and the second most electropositive halogen; trailing behind astatine in both of these categories.
Frequency spectrum	A source of light can have many colors mixed together and in different amounts (intensities.) A rainbow sends the different frequencies in different directions, making them individually visible at different angles. A graph of the intensity plotted against the frequency (showing the amount of each color) is the Frequency spectrum of the light.
Product	A Product is a substance that forms as a result of a biological- or chemical reaction. While the end Product of some chemical reactions may be the result of a relatively rapid reaction, nanoseconds to seconds, chemical equilibria in complex systems may require years or even centuries to be established. For example, equilibria in groundwater systems with multiple components are achieved on timescales of millennia, if ever.
Catalyst	Catalyst is a science centre and museum devoted to the chemical industry. Its full title is Catalyst Science Discovery Centre. It is located in Widnes, Cheshire, in the north west of England, and situated on the north bank of the River Mersey (grid reference SJ512841.)
Principle	A Principle is a comprehensive and fundamental law, doctrine, or assumption. It can be a rule or code of conduct. It can be a law or fact of nature underlying the working of an artificial device.
Yield	Nuclear fission splits a heavy nucleus such as uranium or plutonium into two lighter nuclei, which are called fission products. Yield refers to the fraction of a fission product produced per fission.

Chapter 14. Chemical Equilibrium

Chapter 14. Chemical Equilibrium

Yield can be broken down by:

1. Individual isotope
2. Chemical element spanning several isotopes of different mass number but same atomic number.
3. Nuclei of a given mass number regardless of atomic number. Known as "chain Yield" because it represents a decay chain of beta decay.

Isotope and element yields will change as the fission products undergo beta decay, while chain yields do not change after completion of neutron emission by a few neutron-rich initial fission products (delayed neutrons), with halflife measured in seconds.

A few isotopes can be produced directly by fission, but not by beta decay because the would-be precursor with atomic number one greater is stable and does not decay.

Endothermic	In thermodynamics, the word Endothermic "within-heating" describes a process or reaction that absorbs energy in the form of heat. Its etymology stems from the Greek prefix endo-, meaning "inside" and the Greek suffix -thermic, meaning "to heat". The opposite of an Endothermic process is an exothermic process, one that releases energy in the form of heat.
Graphite	Graphite " href="/wiki/Cleavage_(crystal)">loose interlamellar coupling between sheets in the structure, in fact in a vacuum environment (such as in technologies for use in space), Graphite was found to be a very poor lubricant. This fact led to the discovery that Graphite"s lubricity is due to adsorbed air and water between the layers, unlike other layered dry lubricants such as molybdenum disulfide. Recent studies suggest that an effect called superlubricity can also account for this effect.
Contact process	The Contact process is the current method of producing sulfuric acid in the high concentrations needed for industrial processes. Vanadium(V) oxide (vanadium pentoxide) is the catalyst employed. This process was patented in 1831 by the British vinegar merchant Peregrine Phillips.
Haber process	The Haber process is the nitrogen fixation reaction of nitrogen gas and hydrogen gas, over an enriched iron catalyst, to produce ammonia. The Haber process is important because ammonia is difficult to produce on an industrial scale, and the fertilizer generated from the ammonia is responsible for sustaining one-third of the Earth"s population. Despite the fact that 78.1% of the air we breathe is nitrogen, the gas is relatively unreactive because nitrogen molecules are held together by strong triple bonds.

Chapter 14. Chemical Equilibrium

Chapter 14. Chemical Equilibrium

Oxygen	Oxygen and -γενî®ς (-genÄ"s) (producer, literally begetter) is the element with atomic number 8 and represented by the symbol O. It is a member of the chalcogen group on the periodic table, and is a highly reactive nonmetallic period 2 element that readily forms compounds (notably oxides) with almost all other elements. At standard temperature and pressure two atoms of the element bind to form di Oxygen , a colorless, odorless, tasteless diatomic gas with the formula O_2. Oxygen is the third most abundant element in the universe by mass after hydrogen and helium and the most abundant element by mass in the Earth"s crust.
Platinum	Platinum is a chemical element with the chemical symbol Pt and an atomic number of 78." It is in Group 10 of the periodic table of elements. A dense, malleable, ductile, precious, gray-white transition metal, Platinum is resistant to corrosion and occurs in some nickel and copper ores along with some native deposits. Platinum is used in jewelry, laboratory equipment, electrical contacts and electrodes, Platinum resistance thermometers, dentistry equipment, and catalytic converters.
Sulfur	Sulfur or sulphur is the chemical element that has the atomic number 16. It is denoted with the symbol S. It is an abundant, multivalent non-metal. Sulfur in its native form is a yellow crystalline solid.
Catalysis	Catalysis is the process in which the rate of a chemical reaction is either increased or decreased by means of a chemical substance known as a catalyst. Unlike other reagents that participate in the chemical reaction, a catalyst is not consumed by the reaction itself. The catalyst may participate in multiple chemical transformations.
Mercury	There are seven isotopes of mercury with Hg-202 being the most abundant (29.86%.) The longest-lived radioisotopes are ^{194}Hg with a half-life of 444 years, and ^{203}Hg with a half-life of 46.612 days. Most of the remaining radioisotopes have half-lives that are less than a day.
Nitric acid	Nitric acid, also known as aqua fortis and spirit of nitre, is a highly corrosive and toxic strong acid that can cause severe burns. Colorless when pure, older samples tend to acquire a yellow cast due to the accumulation of oxides of nitrogen. If the solution contains more than 86% Nitric acid, it is referred to as fuming Nitric acid.
Ostwald process	The Ostwald process is a chemical process for producing nitric acid, which was developed by Wilhelm Ostwald (patented 1902.) It is a mainstay of the modern chemical industry. Historically and practically it is closely associated with the Haber process, which provides the requisite raw material, ammonia.

Chapter 14. Chemical Equilibrium

Chapter 14. Chemical Equilibrium

Acid | An Acid is traditionally considered any chemical compound that, when dissolved in water, gives a solution with a hydrogen ion activity greater than in pure water, i.e. a pH less than 7.0. That approximates the modern definition of Johannes Nicolaus Brønsted and Martin Lowry, who independently defined an Acid as a compound which donates a hydrogen ion (H^+) to another compound (called a base.) Common examples include acetic Acid and sulfuric Acid (used in car batteries.)

Chapter 14. Chemical Equilibrium

Chapter 15. Acids and Bases

Acid-base reaction	An Acid-base reaction is a chemical reaction that occurs between an acid and a base. Several concepts that provide alternative definitions for the reaction mechanisms involved and their application in solving related problems exist. Despite several differences in definitions, their importance becomes apparent as different methods of analysis when applied to Acid-base reaction s for gaseous or liquid species, or when acid or base character may be somewhat less apparent.
Ammonia	Ammonia is a compound of nitrogen and hydrogen with the formula NH_3. It is normally encountered as a gas with a characteristic pungent odor. Ammonia contributes significantly to the nutritional needs of terrestrial organisms by serving as a precursor to foodstuffs and fertilizers.
Hydrochloric acid	Hydrochloric acid is the solution of hydrogen chloride (HCl) in water. It is a highly corrosive, strong mineral acid and has major industrial uses. It is found naturally in gastric acid.
Hydrogen	Hydrogen is the chemical element with atomic number 1. It is represented by the symbol H. At standard temperature and pressure, Hydrogen is a colorless, odorless, nonmetallic, tasteless, highly flammable diatomic gas with the molecular formula H_2. With an atomic weight of 1.007 94 u, Hydrogen is the lightest element.
Litmus	Litmus is a water-soluble mixture of different dyes extracted from lichens, especially Roccella tinctoria. The mixture has CAS number 1393-92-6. It is often absorbed onto filter paper.
Acid	An Acid is traditionally considered any chemical compound that, when dissolved in water, gives a solution with a hydrogen ion activity greater than in pure water, i.e. a pH less than 7.0. That approximates the modern definition of Johannes Nicolaus Brønsted and Martin Lowry, who independently defined an Acid as a compound which donates a hydrogen ion (H^+) to another compound (called a base.) Common examples include acetic Acid and sulfuric Acid (used in car batteries.)
Aqueous solution	An Aqueous solution is a solution in which the solvent is water. It is usually shown in chemical equations by appending (aq) to the relevant formula. The word aqueous means pertaining to, related to, similar to, or dissolved in water.
Base	In chemistry, a Base is most commonly thought of as an aqueous substance that can accept hydrogen ions. A Base is also often referred to as an alkali if OH^- ions are involved. This refers to the Brønsted-Lowry theory of acids and bases.
Solution	In chemistry, a Solution is a homogeneous mixture composed of two or more substances. In such a mixture, a solute is dissolved in another substance, known as a solvent. Gases may dissolve in liquids, for example, carbon dioxide or oxygen in water.

Chapter 15. Acids and Bases

Chapter 15. Acids and Bases

Group	In chemistry, a Group is a vertical column in the periodic table of the chemical elements. The name family is derived from the fact that the elements share similar characteristics and traits, just as members of any human family would. There are 18 groups in the standard periodic table.
Hydronium	In chemistry, Hydronium is the common name for the aqueous cation H_3O^+, the simplest type of oxonium ion, produced by protonation of water. It is the positive ion present when an Arrhenius acid is dissolved in water, as Arrhenius acid molecules in solution give up a proton (a positive hydrogen ion, H^+) to the surrounding water molecules (H_2O.) It is the presence of Hydronium ion relative to hydroxide that determines a solution"s pH. The molecules in pure water auto-dissociate into Hydronium and hydroxide ions in the following equillibrium: $2H_2O \; OH^- + H_3O^+$ In pure water, there is an equal number of hydroxide and Hydronium ions, and the pH is perfectly neutral (7.0.)
Neutralization	In chemistry, Neutralization, or neutralisation , is a chemical reaction in which an acid and a base or alkali (soluble base) react to produce salt and water (H_2O.) During the process, hydrogen ions H^+ (a bare proton) from the acid (proton donor) or a hydronium ion H_3O^+ and hydroxide ions OH^- or oxide ions O^{2-} from the base (proton acceptor) react together to form a water molecule H_2O. In the process, a salt is also formed when the anion from acid and the cation from base react together. Neutralization reactions are generally classified as exothermic since heat is released into the surroundings.
Perchloric acid	Perchloric acid, $HClO_4$, is an oxoacid of chlorine and is a colorless liquid soluble in water. It is a strong acid comparable in strength to sulfuric and nitric acids. It is useful for preparing perchlorate salts, but it is also dangerously corrosive and readily forms explosive mixtures.
Sodium	Sodium is a metallic element with a symbol Na and atomic number 11. It is a soft, silvery-white, highly reactive metal and is a member of the alkali metals within "group 1" (formerly known as "group IA".) It has only one stable isotope, ^{23}Na.
Sodium hydroxide	Sodium hydroxide is a caustic metallic base. Sodium hydroxide forms a strong alkaline solution when dissolved in a solvent such as water. However, only the hydroxide ion is basic.
Strong acid	A Strong acid is an acid that dissociates completely in an aqueous solution (not in the case of sulfuric acid as it is diprotic), or in other terms, with a $pK_a < -1.74$. This generally means that in aqueous solution at standard temperature and pressure, the concentration of hydronium ions is equal to the concentration of Strong acid introduced to the solution. While Strong acid s are generally assumed to be the most corrosive, this is not always true.

Chapter 15. Acids and Bases

Chapter 15. Acids and Bases

Colloid	A Colloid is a type of chemical mixture where one substance is dispersed evenly throughout another. The particles of the dispersed substance are only suspended in the mixture, unlike a solution, where they are completely dissolved within. This occurs because the particles in a Colloid are larger than in a solution - small enough to be dispersed evenly and maintain a homogenous appearance, but large enough to scatter light and not dissolve.
Hydroxide	In chemistry, Hydroxide is the name for the diatomic anion OH^-, consisting of oxygen and hydrogen atoms, usually derived from the dissociation of a base. It is one of the simplest diatomic ions known. Inorganic compounds that contain the hydroxyl group are referred to as Hydroxide s.
Ion	An Ion is an atom or molecule where the total number of electrons is not equal to the total number of protons, giving it a net positive or negative electrical charge. Since protons are positively charged and electrons are negatively charged, if there are more electrons than protons, the atom or molecule will be negatively charged. This is called an an Ion , from the Greek á¼€vÎ¬ , meaning "up".
Water	Water is the most abundant molecule on Earth"s surface, constituting about 75% of the Earth"s surface. In nature it exists in liquid, solid, and gaseous states. It is in dynamic equilibrium between the liquid and gas states at standard temperature and pressure.
Covalent bond	A Covalent bond is a form of chemical bonding that is characterized by the sharing of pairs of electrons between atoms, or between atoms and other Covalent bond s. In short, attraction-to-repulsion stability that forms between atoms when they share electrons is known as Covalent bond ing. Covalent bond ing includes many kinds of interaction, including σ-bonding, π-bonding, metal to non-metal bonding, agostic interactions, and three-center two-electron bonds.
Lewis acid	A Lewis acid is a chemical compound, A, that can accept a pair of electrons from a Lewis base, B, that acts as an electron-pair donor, forming an adduct, AB. $$A + :B \rightarrow A\text{--}B$$ Gilbert N. Lewis proposed this definition, which is based on chemical bonding theory, in 1923. Brønsted-Lowry acid-base theory was published in the same year. The two theories are distinct but complementary to each other as a Lewis base is also a Brønsted-Lowry base, but a Lewis acid need not be a Brønsted-Lowry acid.
Oxidation number	The Oxidation number of a central atom in a coordination compound is the charge that it would have if all the ligands were removed along with the electron pairs that were shared with the central atom. It is used in the nomenclature of inorganic compounds. It is represented by a Roman numeral; the plus sign is omitted for positive Oxidation number s.

Chapter 15. Acids and Bases

Chapter 15. Acids and Bases

Complex	In chemistry, a Complex is a structure consisting of a central atom or molecule connected to surrounding atoms or molecules. Originally, a Complex implied a reversible association of molecules, atoms, or ions through weak chemical bonds. As applied to coordination chemistry, this meaning has evolved.
Boron	Boron is the chemical element with atomic number 5 and the chemical symbol B. Boron is a trivalent metalloid element which occurs abundantly in the evaporite ores borax and ulexite. Several allotropes of Boron exist; amorphous Boron is a brown powder, though crystalline Boron is black, extremely hard , and a poor conductor at room temperature. Elemental Boron is used as a dopant in the semiconductor industry, while Boron compounds play important roles as light structural materials, insecticides and preservatives, and reagents for chemical synthesis.
Cadmium	Cadmium is a chemical element with the symbol Cd and atomic number 48. The soft, bluish-white transition metal is chemically similar to the two other metals in group 12 , zinc and mercury. Similar to zinc it prefers oxidation state +2 in most of its compounds and similar to mercury it shows a low melting point for a transition metal.
Cobalt	Cobalt is a hard, lustrous, grey metal, a chemical element with symbol Co and atomic number 27. Although Cobalt based colors and pigments have been used since ancient times for making jewelry and paints, and miners have long used the name kobold ore for some minerals, the free metallic Cobalt was not prepared and discovered until 1735 by Georg Brandt.
Ionization	Ionization is the physical process of converting an atom or molecule into an ion by adding or removing charged particles such as electrons or other ions. This is often confused with dissociation (chemistry.) The process works slightly differently depending on whether an ion with a positive or a negative electric charge is being produced.
Hydrogen iodide	Hydrogen iodide is a diatomic molecule. Aqueous solutions of Hydrogen iodide are known as iohydroic acid or hydriodic acid, a strong acid. Hydrogen iodide and hydroiodic acid are, however, different in that the former is a gas under standard conditions; whereas, the other is an aqueous solution of said gas.
Frequency spectrum	A source of light can have many colors mixed together and in different amounts (intensities.) A rainbow sends the different frequencies in different directions, making them individually visible at different angles. A graph of the intensity plotted against the frequency (showing the amount of each color) is the Frequency spectrum of the light.

Chapter 15. Acids and Bases

Chapter 15. Acids and Bases

Product	A Product is a substance that forms as a result of a biological- or chemical reaction. While the end Product of some chemical reactions may be the result of a relatively rapid reaction, nanoseconds to seconds, chemical equilibria in complex systems may require years or even centuries to be established. For example, equilibria in groundwater systems with multiple components are achieved on timescales of millennia, if ever.
Yield	Nuclear fission splits a heavy nucleus such as uranium or plutonium into two lighter nuclei, which are called fission products. Yield refers to the fraction of a fission product produced per fission. Yield can be broken down by: 1. Individual isotope 2. Chemical element spanning several isotopes of different mass number but same atomic number. 3. Nuclei of a given mass number regardless of atomic number. Known as "chain Yield" because it represents a decay chain of beta decay. Isotope and element yields will change as the fission products undergo beta decay, while chain yields do not change after completion of neutron emission by a few neutron-rich initial fission products (delayed neutrons), with halflife measured in seconds. A few isotopes can be produced directly by fission, but not by beta decay because the would-be precursor with atomic number one greater is stable and does not decay.
Acid rain	Acid rain is rain or any other form of precipitation that is unusually acidic. It has harmful effects on plants, aquatic animals, and infrastructure. Acid rain is mostly caused by human emissions of sulfur and nitrogen compounds which react in the atmosphere to produce acids.
Binary acids	Binary acids are certain molecular compounds in which hydrogen is combined with a second nonmetallic element. Examples: CH_4, NH_3, H_2O, HF, HCl, HBr, HI Their strengths depend on the solvation of the initial acid, the H-X bond energy, the electron affinity energy of X, and the solvation energy of X. Observed trends in acidity correlate with bond energies, the weaker the H-X bond, the stronger the acid. For example, there is a weak bond between hydrogen and iodine in hydroiodic acid, making it a very strong acid.
Electron	The Electron is a subatomic particle that carries a negative electric charge. It has no known substructure and is believed to be a point particle. An Electron has a mass that is approximately 1836 times less than that of the proton.
Electron density	Electron density is the measure of the probability of an electron being present at a specific location.

Chapter 15. Acids and Bases

Chapter 15. Acids and Bases

	In molecules, regions of Electron density are usually found around the atom, and its bonds. In de-localized or conjugated systems, such as phenol, benzene and compounds such as hemoglobin and chlorophyll, the Electron density covers an entire region, i.e., in benzene they are found above and below the planar ring.
Oxoacid	An Oxoacid is an acid which contains oxygen. More specifically, it is an acid which: 1. contains oxygen; 2. contains at least one other element; 3. has at least one hydrogen atom bound to oxygen; and 4. forms an ion by the loss of one or more protons. The name oxyacid is sometimes used, although this is not recommended. Generally, Oxoacid s are simply polyatomic ions with a hydrogen cation. Under Lavoisier"s original theory, all acids contained oxygen, which was named from the Greek οξυς (acid, sharp) and γεινομαι (geinomai) (engender.)
Autoionization	Autoionization is a process by which atoms or molecules spontaneously transition from an electrically neutral state to a lower-energy ionized state. Atoms can autoionize when the two valence electrons are both excited and, through electrostatic interactions, one of the electrons is ejected, ionizing the atom. Molecules can experience vibrational Autoionization.
Self-ionization of water	The Self-ionization of water is the chemical reaction in which two water molecules react to produce a hydronium ion (H_3O^+) and a hydroxide ion (OH^-): $2\ H_2O\ (l)\ H_3O^+\ (aq) + OH^-\ (aq)$ It is an example of autoprotolysis, and relies on the amphoteric nature of water. Water, however pure, is not a simple collection of H_2O molecules. Even in "pure" water, sensitive equipment can detect a very slight electrical conductivity of 0.055 ÂµSÂ·cm^{-1}.
Catalyst	Catalyst is a science centre and museum devoted to the chemical industry. Its full title is Catalyst Science Discovery Centre. It is located in Widnes, Cheshire, in the north west of England, and situated on the north bank of the River Mersey (grid reference SJ512841.)
Principle	A Principle is a comprehensive and fundamental law, doctrine, or assumption. It can be a rule or code of conduct. It can be a law or fact of nature underlying the working of an artificial device.
Calcium	Calcium is the chemical element with the symbol Ca and atomic number 20. It has an atomic mass of 40.078 amu. Calcium is a soft grey alkaline earth metal, and is the fifth most abundant element by mass in the Earth"s crust.
Nitric acid	Nitric acid, also known as aqua fortis and spirit of nitre, is a highly corrosive and toxic strong acid that can cause severe burns.

Chapter 15. Acids and Bases

Chapter 15. Acids and Bases

	Colorless when pure, older samples tend to acquire a yellow cast due to the accumulation of oxides of nitrogen. If the solution contains more than 86% Nitric acid, it is referred to as fuming Nitric acid.
Buffer solution	A Buffer solution is an aqueous solution consisting of a mixture of a weak acid and its conjugate base or a weak base and its conjugate acid. It has the property that the pH of the solution changes very little when a small amount of acid or base is added to it. Buffer solution s are used as a means of keeping pH at a nearly constant value in a wide variety of chemical applications.
Concentration	In chemistry, Concentration is the measure of how much of a given substance there is mixed with another substance. This can apply to any sort of chemical mixture, but most frequently the concept is limited to homogeneous solutions, where it refers to the amount of solute in the solvent. To concentrate a solution, one must add more solute, or reduce the amount of solvent (for instance, by selective evaporation.)
PH meter	A PH meter is an electronic instrument used to measure the pH (acidity or alkalinity) of a liquid (though special probes are sometimes used to measure the pH of semi-solid substances.) A typical PH meter consists of a special measuring probe (a glass electrode) connected to an electronic meter that measures and displays the pH reading. The pH probe measures pH as the activity of hydrogen ions surrounding a thin-walled glass bulb at its tip.
Electrolysis	In chemistry and manufacturing, Electrolysis is a method of using an electric current to drive an otherwise non-spontaneous chemical reaction. Electrolysis is commercially highly important as a stage in the separation of elements from naturally-occurring sources such as ores using an electrolytic cell. - 1800 - William Nicholson and Johann Ritter decomposed water into hydrogen and oxygen. - 1807 - Potassium, Sodium, Barium, Calcium and Magnesium were discovered by Sir Humphry Davy using Electrolysis. - 1886 - Fluorine was discovered by Henri Moissan using Electrolysis. - 1886 - Hall-Héroult process developed for making aluminium - 1890 - Castner-Kellner process developed for making sodium hydroxide Electrolysis is the passage of an electric current through an ionic substance that is either molten or dissolved in a suitable solvent, resulting in chemical reactions at the electrodes and separation of materials.

Chapter 15. Acids and Bases

Chapter 15. Acids and Bases

The main components required to achieve Electrolysis are:

- A liquid containing mobile ions - an electrolyte
- An external source of direct electric current
- Two solid rods or plates known as electrodes

The components perform the following roles in the Electrolysis process:

- The mobile ions are the carriers of electrical current in the liquid (electrolyte.) If the ions are not mobile, as in a solid salt then Electrolysis cannot occur.

Chapter 15. Acids and Bases

Chapter 16. Acid—Base Equilibria

Hydronium

In chemistry, Hydronium is the common name for the aqueous cation H_3O^+, the simplest type of oxonium ion, produced by protonation of water. It is the positive ion present when an Arrhenius acid is dissolved in water, as Arrhenius acid molecules in solution give up a proton (a positive hydrogen ion, H^+) to the surrounding water molecules (H_2O.)

It is the presence of Hydronium ion relative to hydroxide that determines a solution"s pH. The molecules in pure water auto-dissociate into Hydronium and hydroxide ions in the following equillibrium:

$$2H_2O \; OH^- + H_3O^+$$

In pure water, there is an equal number of hydroxide and Hydronium ions, and the pH is perfectly neutral (7.0.)

Ionization

Ionization is the physical process of converting an atom or molecule into an ion by adding or removing charged particles such as electrons or other ions. This is often confused with dissociation (chemistry.)

The process works slightly differently depending on whether an ion with a positive or a negative electric charge is being produced.

Nickel

Nickel is a chemical element, with the chemical symbol Ni and atomic number 28. It is a silvery-white lustrous metal with a slight golden tinge. It is one of the four ferromagnetic elements at about room temperature.

Sodium

Sodium is a metallic element with a symbol Na and atomic number 11. It is a soft, silvery-white, highly reactive metal and is a member of the alkali metals within "group 1" (formerly known as "group IA".) It has only one stable isotope, ^{23}Na.

Solution

In chemistry, a Solution is a homogeneous mixture composed of two or more substances. In such a mixture, a solute is dissolved in another substance, known as a solvent. Gases may dissolve in liquids, for example, carbon dioxide or oxygen in water.

Acid

An Acid is traditionally considered any chemical compound that, when dissolved in water, gives a solution with a hydrogen ion activity greater than in pure water, i.e. a pH less than 7.0. That approximates the modern definition of Johannes Nicolaus Brønsted and Martin Lowry, who independently defined an Acid as a compound which donates a hydrogen ion (H^+) to another compound (called a base.) Common examples include acetic Acid and sulfuric Acid (used in car batteries.)

Base

In chemistry, a Base is most commonly thought of as an aqueous substance that can accept hydrogen ions. A Base is also often referred to as an alkali if OH^- ions are involved. This refers to the Brønsted-Lowry theory of acids and bases.

Chapter 16. Acid—Base Equilibria

Chapter 16. Acid—Base Equilibria

Concentration — In chemistry, Concentration is the measure of how much of a given substance there is mixed with another substance. This can apply to any sort of chemical mixture, but most frequently the concept is limited to homogeneous solutions, where it refers to the amount of solute in the solvent. To concentrate a solution, one must add more solute, or reduce the amount of solvent (for instance, by selective evaporation.)

Ion — An Ion is an atom or molecule where the total number of electrons is not equal to the total number of protons, giving it a net positive or negative electrical charge.
Since protons are positively charged and electrons are negatively charged, if there are more electrons than protons, the atom or molecule will be negatively charged. This is called an an Ion , from the Greek á¼€vÎ¬ , meaning "up".

Degree of ionization — The Degree of ionization refers to the proportion of neutral particles such as those in a gas or aqueous solution, that are ionized into charged particles. A low Degree of ionization is sometimes called partially ionized, and a very high Degree of ionization as fully ionized.
Ionization refers to the process whereby an atom or molecule loses an electron, resulting in two oppositely charged particles, (1) a negatively charged electron and (2) a positively charged ion.

Buffer solution — A Buffer solution is an aqueous solution consisting of a mixture of a weak acid and its conjugate base or a weak base and its conjugate acid. It has the property that the pH of the solution changes very little when a small amount of acid or base is added to it. Buffer solution s are used as a means of keeping pH at a nearly constant value in a wide variety of chemical applications.

Carbonic acid — Carbonic acid (ancient name acid of air or aerial acid) has the formula H_2CO_3. It is also a name sometimes given to solutions of carbon dioxide in water, which contain small amounts of H_2CO_3.
The salts of Carbonic acid s are called bicarbonates (or hydrogen carbonates) and carbonates.

Diprotic acid — A Diprotic acid is an acid such as H_2SO_4 (sulfuric acid) that happens to contain within its molecular structure two hydrogen atoms capable of dissociating (i.e. ionizable) in water. The complete dissociation of Diprotic acid s is of the same form as sulfuric acid:

$$H_2SO_4 \rightarrow H^+(aq) + HSO_4^-(aq) \quad K_a = 1 \times 10^3$$

$$HSO_4^- \rightarrow H^+(aq) + SO_4^{2-}(aq) \quad K_a = 1 \times 10^{-2}$$

The dissociation does not happen all at once due to the two stages of dissociation having different K_a values. The first dissociation will, in the case of sulfuric acid, occur completely, but the second one will not.

Acid rain — Acid rain is rain or any other form of precipitation that is unusually acidic. It has harmful effects on plants, aquatic animals, and infrastructure. Acid rain is mostly caused by human emissions of sulfur and nitrogen compounds which react in the atmosphere to produce acids.

Chapter 16. Acid—Base Equilibria

Chapter 16. Acid—Base Equilibria

Ascorbic acid	Ascorbic acid is a sugar acid with antioxidant properties. Its appearance is white to light-yellow crystals or powder, and it is water-soluble. One form of Ascorbic acid is commonly known as vitamin C. In 1937 the Nobel Prize for chemistry was awarded to Walter Haworth for his work in determining the structure of Ascorbic acid, and the prize for Physiology or Medicine that year went to Albert Szent-Györgyi for his studies of the biological functions of L-Ascorbic acid.
Calcium	Calcium is the chemical element with the symbol Ca and atomic number 20. It has an atomic mass of 40.078 amu. Calcium is a soft grey alkaline earth metal, and is the fifth most abundant element by mass in the Earth"s crust.
Hydrochloric acid	Hydrochloric acid is the solution of hydrogen chloride (HCl) in water. It is a highly corrosive, strong mineral acid and has major industrial uses. It is found naturally in gastric acid.
Nitric acid	Nitric acid, also known as aqua fortis and spirit of nitre, is a highly corrosive and toxic strong acid that can cause severe burns. Colorless when pure, older samples tend to acquire a yellow cast due to the accumulation of oxides of nitrogen. If the solution contains more than 86% Nitric acid, it is referred to as fuming Nitric acid.
Ammonia	Ammonia is a compound of nitrogen and hydrogen with the formula NH_3. It is normally encountered as a gas with a characteristic pungent odor. Ammonia contributes significantly to the nutritional needs of terrestrial organisms by serving as a precursor to foodstuffs and fertilizers.
Aqueous solution	An Aqueous solution is a solution in which the solvent is water. It is usually shown in chemical equations by appending (aq) to the relevant formula. The word aqueous means pertaining to, related to, similar to, or dissolved in water.
Cadmium	Cadmium is a chemical element with the symbol Cd and atomic number 48. The soft, bluish-white transition metal is chemically similar to the two other metals in group 12 , zinc and mercury. Similar to zinc it prefers oxidation state +2 in most of its compounds and similar to mercury it shows a low melting point for a transition metal.
Salt	Salt is a dietary mineral composed primarily of sodium chloride that is essential for animal life, but toxic to most land plants. Salt flavor is one of the basic tastes, an important preservative and a popular food seasoning.
Neutralization	In chemistry, Neutralization, or neutralisation , is a chemical reaction in which an acid and a base or alkali (soluble base) react to produce salt and water (H_2O.) During the process, hydrogen ions H^+ (a bare proton) from the acid (proton donor) or a hydronium ion H_3O^+ and hydroxide ions OH^- or oxide ions O^{2-} from the base (proton acceptor) react together to form a water molecule H_2O. In the process, a salt is also formed when the anion from acid and the cation from base react together.

Chapter 16. Acid—Base Equilibria

Chapter 16. Acid—Base Equilibria

Neutralization reactions are generally classified as exothermic since heat is released into the surroundings.

Hydrogen

Hydrogen is the chemical element with atomic number 1. It is represented by the symbol H. At standard temperature and pressure, Hydrogen is a colorless, odorless, nonmetallic, tasteless, highly flammable diatomic gas with the molecular formula H_2. With an atomic weight of 1.007 94 u, Hydrogen is the lightest element.

Hydrolysis

Hydrolysis is a chemical reaction during which one or more water molecules are split into hydrogen and hydroxide ions, which may go on to participate in further reactions. It is the type of reaction that is used to break down certain polymers, especially those made by step-growth polymerization. Such polymer degradation is usually catalysed by either acid e.g. concentrated sulfuric acid (H_2SO_4) or alkali e.g. sodium hydroxide (NaOH) attack, often increasing with their strength or pH.

Hydrolysis is distinct from hydration, where the hydrated molecule does not "lyse" (break into two new compounds.)

Metal

In chemistry, a Metal is an element, compound, or alloy characterized by high electrical conductivity. In a Metal atoms readily lose electrons to form positive ions ; those ions are surrounded by delocalized electrons, which are responsible for the conductivity. The thus produced solid is held by electrostatic interactions between the ions and the electron cloud, which are called Metal lic bonds.

Potassium

Potassium is a chemical element. It has the symbol K , atomic number 19, and atomic mass 39.0983. Potassium was first isolated from potash.

Strong acid

A Strong acid is an acid that dissociates completely in an aqueous solution (not in the case of sulfuric acid as it is diprotic), or in other terms, with a $pK_a < -1.74$. This generally means that in aqueous solution at standard temperature and pressure, the concentration of hydronium ions is equal to the concentration of Strong acid introduced to the solution. While Strong acid s are generally assumed to be the most corrosive, this is not always true.

Complex

In chemistry, a Complex is a structure consisting of a central atom or molecule connected to surrounding atoms or molecules. Originally, a Complex implied a reversible association of molecules, atoms, or ions through weak chemical bonds. As applied to coordination chemistry, this meaning has evolved.

Sodium fluoride

Sodium fluoride is the inorganic compound with the formula NaF. This colorless solid is a source of the fluoride ion in diverse applications. Sodium fluoride is less expensive and less hygroscopic than the related salt potassium fluoride.

Sodium fluoride is an ionic compound, dissolving to give separated Na^+ and F^- ions.

Chapter 16. Acid—Base Equilibria

Chapter 16. Acid—Base Equilibria

Zinc	Zinc is a metallic chemical element with the symbol Zn and atomic number 30. It is a first-row transition metal in group 12 of the periodic table. Zinc is chemically similar to magnesium because its ion is of similar size and its only common oxidation state is +2.
Common-ion effect	The Common-ion effect is a term used to describe the effect on a solution of two dissolved solutes that contain the same ion. The presence of a common ion suppresses the ionization of a weak acid or a weak base. - If both sodium acetate and acetic acid are dissolved in the same solution they both dissociate and ionize to produce acetate ions. Sodium acetate is a strong electrolyte so it dissociates completely in solution. Acetic acid is a weak acid so it only ionizes slightly. According to Le Chatelier"s principle, the addition of acetate ions from sodium acetate will suppress the ionization of acetic acid and shift its equilibrium to the left. Thus the percent dissociation of the acetic acid will decrease and the pH of the solution will increase. The ionization of an acid or a base is limited by the presence of its conjugate base or acid. $$NaC_2H_3O_{2(s)} \rightarrow Na^+_{(aq)} + C_2H_3O_2^-_{(aq)}$$ $$HC_2H_3O_{2(l)} \leftrightarrow H^+_{(aq)} + C_2H_3O_2^-_{(aq)}$$ This will decrease the hydrogen ion concentration and thus the common-ion solution will be less acidic than a solution containing only acetic acid. - A practical example used very widely in areas drawing drinking water from chalk or limestone aquifers is the addition of sodium carbonate to the raw water to precipitate out hardness salts. In this case, as part of the water treatment process, the highly soluble sodium carbonate salt precipitates out the very sparingly soluble calcium carbonate reducing the overall hardness. The very pure and finely divided precipitate of calcium carbonate that is generated is a valuable by-product used in the manufacture of toothpaste.
Catalyst	Catalyst is a science centre and museum devoted to the chemical industry. Its full title is Catalyst Science Discovery Centre. It is located in Widnes, Cheshire, in the north west of England, and situated on the north bank of the River Mersey (grid reference SJ512841.)
Principle	A Principle is a comprehensive and fundamental law, doctrine, or assumption. It can be a rule or code of conduct. It can be a law or fact of nature underlying the working of an artificial device.
Polymer	A Polymer is a large molecule composed of repeating structural units typically connected by covalent chemical bonds. While Polymer in popular usage suggests plastic, the term actually refers to a large class of natural and synthetic materials with a variety of properties.

Chapter 16. Acid—Base Equilibria

Chapter 16. Acid—Base Equilibria

	Due to the extraordinary range of properties accessible in Polymer ic materials , they have come to play an essential and ubiquitous role in everyday life - from plastics and elastomers on the one hand to natural bio Polymer s such as DNA and proteins that are essential for life on the other.
Citric acid	Citric acid is a weak organic acid, and it is a natural preservative and is also used to add an acidic taste to foods and soft drinks. In biochemistry, it is important as an intermediate in the Citric acid cycle and therefore occurs in the metabolism of virtually all living things. It can also be used as an environmentally benign cleaning agent.
Sodium hydroxide	Sodium hydroxide is a caustic metallic base. Sodium hydroxide forms a strong alkaline solution when dissolved in a solvent such as water. However, only the hydroxide ion is basic.
Hydroxide	In chemistry, Hydroxide is the name for the diatomic anion OH^-, consisting of oxygen and hydrogen atoms, usually derived from the dissociation of a base. It is one of the simplest diatomic ions known.
	Inorganic compounds that contain the hydroxyl group are referred to as Hydroxide s.

Chapter 16. Acid—Base Equilibria

Chapter 17. Solubility and Complex-Ion Equilibria

Calcium	Calcium is the chemical element with the symbol Ca and atomic number 20. It has an atomic mass of 40.078 amu. Calcium is a soft grey alkaline earth metal, and is the fifth most abundant element by mass in the Earth"s crust.
Ionic compound	In chemistry, an Ionic compound is a chemical compound in which ions are held together in a lattice structure by ionic bonds. Usually, the positively charged portion consists of metal cations and the negatively charged portion is an anion or polyatomic ion. Ions in Ionic compound s are held together by the electrostatic force between oppositely charged bodies.
Lead	Lead is a main-group element with symbol Pb and atomic number 82. Lead is a soft, malleable poor metal, also considered to be one of the heavy metals. Lead has a bluish-white color when freshly cut, but tarnishes to a dull grayish color when exposed to air.
Acid	An Acid is traditionally considered any chemical compound that, when dissolved in water, gives a solution with a hydrogen ion activity greater than in pure water, i.e. a pH less than 7.0. That approximates the modern definition of Johannes Nicolaus Brønsted and Martin Lowry, who independently defined an Acid as a compound which donates a hydrogen ion (H^+) to another compound (called a base.) Common examples include acetic Acid and sulfuric Acid (used in car batteries.)
Acid rain	Acid rain is rain or any other form of precipitation that is unusually acidic. It has harmful effects on plants, aquatic animals, and infrastructure. Acid rain is mostly caused by human emissions of sulfur and nitrogen compounds which react in the atmosphere to produce acids.
Ion	An Ion is an atom or molecule where the total number of electrons is not equal to the total number of protons, giving it a net positive or negative electrical charge. Since protons are positively charged and electrons are negatively charged, if there are more electrons than protons, the atom or molecule will be negatively charged. This is called an an Ion , from the Greek á¼€vÎ¬ , meaning "up".
Product	A Product is a substance that forms as a result of a biological- or chemical reaction. While the end Product of some chemical reactions may be the result of a relatively rapid reaction, nanoseconds to seconds, chemical equilibria in complex systems may require years or even centuries to be established. For example, equilibria in groundwater systems with multiple components are achieved on timescales of millennia, if ever.
Mercury	There are seven isotopes of mercury with Hg-202 being the most abundant (29.86%.) The longest-lived radioisotopes are ^{194}Hg with a half-life of 444 years, and ^{203}Hg with a half-life of 46.612 days. Most of the remaining radioisotopes have half-lives that are less than a day.
Silver	Silver is a chemical element with the chemical symbol Ag and atomic number 47. A soft, white, lustrous transition metal, it has the highest electrical conductivity of any element and the highest thermal conductivity of any metal. The metal occurs naturally in its pure, free form , as an alloy with gold (electrum) and other metals, and in minerals such as argentite and chlorargyrite.

Chapter 17. Solubility and Complex-Ion Equilibria

Chapter 17. Solubility and Complex-Ion Equilibria

Calcium fluoride

Calcium fluoride is an insoluble ionic compound of calcium and fluorine. It occurs naturally as the mineral fluorite (also called fluorspar) and as Blue John, and it is the source of most of the world"s fluorine. This insoluble solid adopts a cubic structure wherein calcium is coordinated to eight fluoride anions and each F^- ion is surrounded by four Ca^{2+} ions.

Common-ion effect

The Common-ion effect is a term used to describe the effect on a solution of two dissolved solutes that contain the same ion.
The presence of a common ion suppresses the ionization of a weak acid or a weak base.

- If both sodium acetate and acetic acid are dissolved in the same solution they both dissociate and ionize to produce acetate ions. Sodium acetate is a strong electrolyte so it dissociates completely in solution. Acetic acid is a weak acid so it only ionizes slightly. According to Le Chatelier"s principle, the addition of acetate ions from sodium acetate will suppress the ionization of acetic acid and shift its equilibrium to the left. Thus the percent dissociation of the acetic acid will decrease and the pH of the solution will increase. The ionization of an acid or a base is limited by the presence of its conjugate base or acid.

$$NaC_2H_3O_{2(s)} \rightarrow Na^+_{(aq)} + C_2H_3O_2^-_{(aq)}$$

$$HC_2H_3O_{2(l)} \leftrightarrow H^+_{(aq)} + C_2H_3O_2^-_{(aq)}$$

This will decrease the hydrogen ion concentration and thus the common-ion solution will be less acidic than a solution containing only acetic acid.

- A practical example used very widely in areas drawing drinking water from chalk or limestone aquifers is the addition of sodium carbonate to the raw water to precipitate out hardness salts. In this case, as part of the water treatment process, the highly soluble sodium carbonate salt precipitates out the very sparingly soluble calcium carbonate reducing the overall hardness. The very pure and finely divided precipitate of calcium carbonate that is generated is a valuable by-product used in the manufacture of toothpaste.

Lead paint

Lead paint is paint containing lead, a heavy metal, that is used as pigment, with lead(II) chromate ($PbCrO_4$, "chrome yellow") and lead(II) carbonate($PbCO_3$, "white lead") being the most common. Lead is also added to paint to speed drying, increase durability, retain a fresh appearance, and resist moisture that causes corrosion. Paint with significant lead content is still used in industry and by the military.

Chapter 17. Solubility and Complex-Ion Equilibria

Chapter 17. Solubility and Complex-Ion Equilibria

Catalyst	Catalyst is a science centre and museum devoted to the chemical industry. Its full title is Catalyst Science Discovery Centre. It is located in Widnes, Cheshire, in the north west of England, and situated on the north bank of the River Mersey (grid reference SJ512841.)
Concentration	In chemistry, Concentration is the measure of how much of a given substance there is mixed with another substance. This can apply to any sort of chemical mixture, but most frequently the concept is limited to homogeneous solutions, where it refers to the amount of solute in the solvent. To concentrate a solution, one must add more solute, or reduce the amount of solvent (for instance, by selective evaporation.)
Paint	Paint is any liquid, liquifiable, or mastic composition which after application to a substrate in a thin layer is converted to an opaque solid film. Cave Paint ings drawn with red and yellow ochre, hematite, manganese oxide and charcoal may have been made by early Homo sapiens as long as 40,000 years ago. Ancient Paint ed walls at Denerdera, Egypt, which were exposed for many ages to the open air, still possess a perfect brilliancy of color, as vivid as when they were Paint ed about 2,000 years ago.
Principle	A Principle is a comprehensive and fundamental law, doctrine, or assumption. It can be a rule or code of conduct. It can be a law or fact of nature underlying the working of an artificial device.
Solution	In chemistry, a Solution is a homogeneous mixture composed of two or more substances. In such a mixture, a solute is dissolved in another substance, known as a solvent. Gases may dissolve in liquids, for example, carbon dioxide or oxygen in water.
Precipitation	Precipitation is the formation of a solid in a solution during a chemical reaction. When the reaction occurs, the solid formed is called the precipitate, and the liquid remaining above the solid is called the supernate. Natural methods of Precipitation include settling or sedimentation, where a solid forms over a period of time due to ambient forces like gravity or centrifugation.
Sodium	Sodium is a metallic element with a symbol Na and atomic number 11. It is a soft, silvery-white, highly reactive metal and is a member of the alkali metals within "group 1" (formerly known as "group IA".) It has only one stable isotope, ^{23}Na.
Barium	Barium is a chemical element. It has the symbol Ba, and atomic number 56. Barium is a soft silvery metallic alkaline earth metal.
Seawater	Seawater is water from a sea or ocean. On average, Seawater in the world"s oceans has a salinity of about 3.5%. This means that every 1 kg of Seawater has approximately 35 grams of dissolved salts (mostly, but not entirely, the ions of sodium chloride: Na^+, Cl^-.)

Chapter 17. Solubility and Complex-Ion Equilibria

Chapter 17. Solubility and Complex-Ion Equilibria

Magnesium	Magnesium is a chemical element with the symbol Mg, atomic number 12, atomic weight 24.3050 and common oxidation number +2.
	Magnesium, an alkaline earth metal, is the ninth most abundant element in the universe by mass. The commonness of Magnesium is related to the fact that it is easily built up in supernova stars from a sequential addition of three helium nuclei to carbon .
Potassium	Potassium is a chemical element. It has the symbol K , atomic number 19, and atomic mass 39.0983. Potassium was first isolated from potash.
Buffer solution	A Buffer solution is an aqueous solution consisting of a mixture of a weak acid and its conjugate base or a weak base and its conjugate acid. It has the property that the pH of the solution changes very little when a small amount of acid or base is added to it. Buffer solution s are used as a means of keeping pH at a nearly constant value in a wide variety of chemical applications.
Hydrogen	Hydrogen is the chemical element with atomic number 1. It is represented by the symbol H. At standard temperature and pressure, Hydrogen is a colorless, odorless, nonmetallic, tasteless, highly flammable diatomic gas with the molecular formula H_2. With an atomic weight of 1.007 94 u, Hydrogen is the lightest element.
Hydrogen sulfide	Hydrogen sulfide is the chemical compound with the formula H_2S. This colorless, toxic and flammable gas is partially responsible for the foul odor of rotten eggs and flatulence.
	It often results from the bacterial break down of sulfites in nonorganic matter in the absence of oxygen, such as in swamps and sewers (anaerobic digestion.) It also occurs in volcanic gases, natural gas and some well waters.
Metal	In chemistry, a Metal is an element, compound, or alloy characterized by high electrical conductivity. In a Metal atoms readily lose electrons to form positive ions ; those ions are surrounded by delocalized electrons, which are responsible for the conductivity. The thus produced solid is held by electrostatic interactions between the ions and the electron cloud, which are called Metal lic bonds.
Zinc	Zinc is a metallic chemical element with the symbol Zn and atomic number 30. It is a first-row transition metal in group 12 of the periodic table. Zinc is chemically similar to magnesium because its ion is of similar size and its only common oxidation state is +2.
Boiling	Boiling, a type of phase transition, is the rapid vaporization of a liquid, which typically occurs when a liquid is heated to its Boiling point, the temperature at which the vapor pressure of the liquid is equal to the pressure exerted on the liquid by the surrounding environmental pressure. Thus, a liquid may also boil when the pressure of the surrounding atmosphere is sufficiently reduced, such as the use of a vacuum pump or at high altitudes. Boiling occurs in three characteristic stages, which are nucleate, transition and film Boiling.

Chapter 17. Solubility and Complex-Ion Equilibria

Chapter 17. Solubility and Complex-Ion Equilibria

Complex	In chemistry, a Complex is a structure consisting of a central atom or molecule connected to surrounding atoms or molecules. Originally, a Complex implied a reversible association of molecules, atoms, or ions through weak chemical bonds. As applied to coordination chemistry, this meaning has evolved.
Crystal	A Crystal or Crystal line solid is a solid material whose constituent atoms, molecules, or ions are arranged in an orderly repeating pattern extending in all three spatial dimensions. The scientific study of Crystal s and Crystal formation is Crystal lography. The process of Crystal formation via mechanisms of Crystal growth is called Crystal lization or solidification.
Crystal structure	In mineralogy and crystallography, a Crystal structure is a unique arrangement of atoms in a crystal. A Crystal structure is composed of a motif, a set of atoms arranged in a particular way, and a lattice. Motifs are located upon the points of a lattice, which is an array of points repeating periodically in three dimensions.
Ionization	Ionization is the physical process of converting an atom or molecule into an ion by adding or removing charged particles such as electrons or other ions. This is often confused with dissociation (chemistry.) The process works slightly differently depending on whether an ion with a positive or a negative electric charge is being produced.
Melting	Melting is a physical process that results in the phase change of a substance from a solid to a liquid. The internal energy of a solid substance is increased, typically by the application of heat or pressure, resulting in a rise of its temperature to the Melting point, at which the rigid ordering of molecular entities in the solid breaks down to a less-ordered state and the solid liquefies. An object that has melted completely is molten.
Dissociation	Dissociation in chemistry and biochemistry is a general process in which ionic compounds (complexes, molecules ions usually in a reversible manner. When a Bronsted-Lowry acid is put in water, a covalent bond between an electronegative atom and a hydrogen atom is broken by heterolytic fission, which gives a proton and a negative ion. Dissociation is the opposite of association and recombination.
Ligand	In chemistry, a Ligand is either an atom, ion, or molecule that binds to a central metal to produce a coordination complex. The bonding between the metal and Ligand generally involves formal donation of one or more of the Ligand"s electrons. The metal-Ligand bonding ranges from covalent to more ionic.
Stability constant	A stability constant is an equilibrium constant for the formation of a complex in solution. It is a measure of the strength of the interaction between the reagents that come together to form the complex. There are two main kinds of complex: compounds formed by the interaction of a metal ion with a ligand and supramolecular complexes, such as host-guest complexes and complexes of anions.

Chapter 17. Solubility and Complex-Ion Equilibria

Chapter 17. Solubility and Complex-Ion Equilibria

Atomic mass	The Atomic mass is the mass of an atom, most often expressed in unified Atomic mass units. The Atomic mass may be considered to be the total mass of protons, neutrons and electrons in a single atom (when the atom is motionless.) The Atomic mass is sometimes incorrectly used as a synonym of relative Atomic mass, average Atomic mass and atomic weight; however, these differ subtly from the Atomic mass.
Amphoteric	In chemistry, an amphoteric substance is one that can react as either an acid or base. Many metals (such as zinc, tin, lead, aluminium, and beryllium) and most metalloids have amphoteric oxides or hydroxides. Another class of amphoteric substances are amphiprotic molecules which can either donate or accept a proton.
Chromium	Chromium is a chemical element which has the symbol Cr and atomic number 24. It is a steely-gray, lustrous, hard metal that takes a high polish and has a high melting point. It is also odourless, tasteless, and malleable.
Tin	Tin is a chemical element with the symbol Sn and atomic number 50. It is a main group metal in group 14 of the periodic table. Tin shows chemical similarity to both neighboring group 14 elements, germanium and lead, like the two possible oxidation states +2 and +4.
Hydroxide	In chemistry, Hydroxide is the name for the diatomic anion OH^-, consisting of oxygen and hydrogen atoms, usually derived from the dissociation of a base. It is one of the simplest diatomic ions known. Inorganic compounds that contain the hydroxyl group are referred to as Hydroxide s.

Chapter 17. Solubility and Complex-Ion Equilibria

Chapter 18. Thermodynamics and Equilibrium

Ammonia	Ammonia is a compound of nitrogen and hydrogen with the formula NH_3. It is normally encountered as a gas with a characteristic pungent odor. Ammonia contributes significantly to the nutritional needs of terrestrial organisms by serving as a precursor to foodstuffs and fertilizers.
Carbon	Carbon is the chemical element with symbol C and atomic number 6. As a member of group 14 on the periodic table, it is nonmetallic and tetravalent--making four electrons available to form covalent chemical bonds. There are three naturally occurring isotopes, with ^{12}C and ^{13}C being stable, while ^{14}C is radioactive, decaying with a half-life of about 5730 years.
Enthalpy	In thermodynamics and molecular chemistry, the Enthalpy is a thermodynamic property of a fluid. It can be used to calculate the heat transfer during a quasistatic process taking place in a closed thermodynamic system under constant pressure. Enthalpy H is an arbitrary concept but the Enthalpy change ΔH is more useful because it is equal to the change in the internal energy of the system, plus the work that the system has done on its surroundings.
Internal energy	In thermodynamics, the Internal energy of a thermodynamic system denoted by U is the total of the kinetic energy due to the motion of molecules (translational, rotational, vibrational) and the potential energy associated with the vibrational and electric energy of atoms within molecules or crystals. It includes the energy in all of the chemical bonds, and the energy of the free, conduction electrons in metals. One can also calculate the Internal energy of electromagnetic or blackbody radiation.
Uranium	Uranium is a silvery-white metallic chemical element in the actinide series of the periodic table that has the symbol U and atomic number 92. Besides its 92 protons, a Uranium nucleus can have between 141 and 146 neutrons, with 146 and 143 (U-235) in its most common isotopes. The number of electrons in a Uranium atom is 92, 6 of them valence electrons.
Aqueous solution	An Aqueous solution is a solution in which the solvent is water. It is usually shown in chemical equations by appending (aq) to the relevant formula. The word aqueous means pertaining to, related to, similar to, or dissolved in water.
Entropy	In thermodynamics, Entropy is often associated with the amount of order, disorder, and/or chaos in a thermodynamic system. This stems from Rudolf Clausius" 1862 assertion that any thermodynamic processes always "admits to being reduced to the alteration in some way or another of the arrangement of the constituent parts of the working body" and that internal work associated with these alterations is quantified energetically by a measure of "Entropy" change, according to the following differential expression: $$\int \frac{\delta Q}{T} \geq 0$$ In the years to follow, Ludwig Boltzmann translated these "alterations" into that of a probabilistic view of order and disorder in gas phase molecular systems.

Chapter 18. Thermodynamics and Equilibrium

Chapter 18. Thermodynamics and Equilibrium

In recent years, in some chemistry publications, there has been a shift away from using the terms "order" and "disorder" to that of the concept of energy dispersion to describe Entropy, among other theories.

Molecular model

A Molecular model in this article, is a physical model that represents molecules and their processes. The creation of mathematical models of molecular properties and behaviour is Molecular model ling, and their graphical depiction is molecular graphics, but these topics are closely linked and each uses techniques from the others. In this article, Molecular model will primarily refer to systems containing more than one atom and where nuclear structure is neglected.

Solution

In chemistry, a Solution is a homogeneous mixture composed of two or more substances. In such a mixture, a solute is dissolved in another substance, known as a solvent. Gases may dissolve in liquids, for example, carbon dioxide or oxygen in water.

Hydrochloric acid

Hydrochloric acid is the solution of hydrogen chloride (HCl) in water. It is a highly corrosive, strong mineral acid and has major industrial uses. It is found naturally in gastric acid.

Zinc

Zinc is a metallic chemical element with the symbol Zn and atomic number 30. It is a first-row transition metal in group 12 of the periodic table. Zinc is chemically similar to magnesium because its ion is of similar size and its only common oxidation state is +2.

Acid

An Acid is traditionally considered any chemical compound that, when dissolved in water, gives a solution with a hydrogen ion activity greater than in pure water, i.e. a pH less than 7.0. That approximates the modern definition of Johannes Nicolaus Brønsted and Martin Lowry, who independently defined an Acid as a compound which donates a hydrogen ion (H^+) to another compound (called a base.) Common examples include acetic Acid and sulfuric Acid (used in car batteries.)

Methanation

Methanation is a physical-chemical process to generate Methane from a mixture of various gases out of biomass fermentation or thermo-chemical gasification. The main components are carbon monoxide and hydrogen. The following main process describes the Methanation:

$$CO + 3\,H_2 \rightarrow CH_4 + H_2O$$

This process is used for the generation of biogenous natural gas substitute, which can be fed into the gas grid.Methanation is the reverse reaction of steam methane reforming, which converts methane into synthesis gas.

Chapter 18. Thermodynamics and Equilibrium

Chapter 18. Thermodynamics and Equilibrium

Oxygen	Oxygen and -γενÎ®ς (-genÄ"s) (producer, literally begetter) is the element with atomic number 8 and represented by the symbol O. It is a member of the chalcogen group on the periodic table, and is a highly reactive nonmetallic period 2 element that readily forms compounds (notably oxides) with almost all other elements. At standard temperature and pressure two atoms of the element bind to form di Oxygen , a colorless, odorless, tasteless diatomic gas with the formula O_2. Oxygen is the third most abundant element in the universe by mass after hydrogen and helium and the most abundant element by mass in the Earth"s crust.
Iron	Iron is a chemical element with the symbol Fe and atomic number 26. Iron is a group 8 and period 4 element. Iron and Iron alloys (steels) are by far the most common metals and the most common ferromagnetic materials in everyday use.
Spontaneous process	A Spontaneous process is the time-evolution of a system in which it releases free energy (most often as heat) and moves to a lower, more thermodynamically stable, energy state. The sign convention of changes in free energy follows the general convention for thermodynamic measurements, in which a release of free energy from the system corresponds to a negative change in free energy, but a positive change for the surroundings. A Spontaneous process is capable of proceeding in a given direction, as written or described, without needing to be driven by an outside source of energy.
Phase transition	In thermodynamics, a Phase transition is the transformation of a thermodynamic system from one phase to another. At a Phase transition point, physical properties may undergo abrupt change: for instance, the volume of the two phases may be vastly different as is illustrated by the boiling of liquid water to form steam. The term is most commonly used to describe transitions between solid, liquid and gaseous states of matter, in rare cases including plasma.
Impurities	Impurities are substances inside a confined amount of liquid, gas which differ from the chemical composition of the material or compound. Impurities are either naturally occurring or added during synthesis of a chemical or commercial product. During production, Impurities may be purposely, accidentally, inevitably, or incidentally added into the substance.
Boiling	Boiling, a type of phase transition, is the rapid vaporization of a liquid, which typically occurs when a liquid is heated to its Boiling point, the temperature at which the vapor pressure of the liquid is equal to the pressure exerted on the liquid by the surrounding environmental pressure. Thus, a liquid may also boil when the pressure of the surrounding atmosphere is sufficiently reduced, such as the use of a vacuum pump or at high altitudes. Boiling occurs in three characteristic stages, which are nucleate, transition and film Boiling.

Chapter 18. Thermodynamics and Equilibrium

Chapter 18. Thermodynamics and Equilibrium

Fermentation	Fermentation is the process of deriving energy from the oxidation of organic compounds, such as carbohydrates, using an endogenous electron acceptor, which is usually an organic compound. This is in contrast to cellular respiration, where electrons are donated to an exogenous electron acceptor, such as oxygen, via an electron transport chain. Fermentation does not necessarily have to be carried out in an anaerobic environment.
Freezing	In physical science, Freezing or solidification is the process in which a liquid turns into a solid when cold enough. The Freezing point is the temperature at which this happens. Melting, the process of turning a solid to a liquid, is almost the exact opposite of Freezing.
Combustion	Combustion or burning is a complex sequence of exothermic chemical reactions between a fuel (usually a hydrocarbon) and an oxidant accompanied by the production of heat or both heat and light in the form of either a glow or flames, appearance of light flickering. Direct Combustion by atmospheric oxygen is a reaction mediated by radical intermediates. The conditions for radical production are naturally produced by thermal runaway, where the heat generated by Combustion is necessary to maintain the high temperature necessary for radical production.
Acid-base reaction	An Acid-base reaction is a chemical reaction that occurs between an acid and a base. Several concepts that provide alternative definitions for the reaction mechanisms involved and their application in solving related problems exist. Despite several differences in definitions, their importance becomes apparent as different methods of analysis when applied to Acid-base reaction s for gaseous or liquid species, or when acid or base character may be somewhat less apparent.
Polymer	A Polymer is a large molecule composed of repeating structural units typically connected by covalent chemical bonds. While Polymer in popular usage suggests plastic, the term actually refers to a large class of natural and synthetic materials with a variety of properties. Due to the extraordinary range of properties accessible in Polymer ic materials , they have come to play an essential and ubiquitous role in everyday life - from plastics and elastomers on the one hand to natural bio Polymer s such as DNA and proteins that are essential for life on the other.

Chapter 18. Thermodynamics and Equilibrium

Chapter 19. Electrochemistry

Copper	Copper is a chemical element with the symbol Cu and atomic number 29. It is a ductile metal with very high thermal and electrical conductivity. Pure Copper is rather soft and malleable and a freshly-exposed surface has a pinkish or peachy color.
Electrolysis	In chemistry and manufacturing, Electrolysis is a method of using an electric current to drive an otherwise non-spontaneous chemical reaction. Electrolysis is commercially highly important as a stage in the separation of elements from naturally-occurring sources such as ores using an electrolytic cell. - 1800 - William Nicholson and Johann Ritter decomposed water into hydrogen and oxygen. - 1807 - Potassium, Sodium, Barium, Calcium and Magnesium were discovered by Sir Humphry Davy using Electrolysis. - 1886 - Fluorine was discovered by Henri Moissan using Electrolysis. - 1886 - Hall-Héroult process developed for making aluminium - 1890 - Castner-Kellner process developed for making sodium hydroxide Electrolysis is the passage of an electric current through an ionic substance that is either molten or dissolved in a suitable solvent, resulting in chemical reactions at the electrodes and separation of materials. The main components required to achieve Electrolysis are: - A liquid containing mobile ions - an electrolyte - An external source of direct electric current - Two solid rods or plates known as electrodes The components perform the following roles in the Electrolysis process: - The mobile ions are the carriers of electrical current in the liquid (electrolyte.) If the ions are not mobile, as in a solid salt then Electrolysis cannot occur.
Hydrogen	Hydrogen is the chemical element with atomic number 1. It is represented by the symbol H. At standard temperature and pressure, Hydrogen is a colorless, odorless, nonmetallic, tasteless, highly flammable diatomic gas with the molecular formula H_2. With an atomic weight of 1.007 94 u, Hydrogen is the lightest element.
Hydronium	In chemistry, Hydronium is the common name for the aqueous cation H_3O^+, the simplest type of oxonium ion, produced by protonation of water. It is the positive ion present when an Arrhenius acid is dissolved in water, as Arrhenius acid molecules in solution give up a proton (a positive hydrogen ion, H^+) to the surrounding water molecules (H_2O.)

Chapter 19. Electrochemistry

Chapter 19. Electrochemistry

	It is the presence of Hydronium ion relative to hydroxide that determines a solution"s pH. The molecules in pure water auto-dissociate into Hydronium and hydroxide ions in the following equillibrium: $$2H_2O \rightleftharpoons OH^- + H_3O^+$$ In pure water, there is an equal number of hydroxide and Hydronium ions, and the pH is perfectly neutral (7.0.)
Zinc	Zinc is a metallic chemical element with the symbol Zn and atomic number 30. It is a first-row transition metal in group 12 of the periodic table. Zinc is chemically similar to magnesium because its ion is of similar size and its only common oxidation state is +2.
Ion	An Ion is an atom or molecule where the total number of electrons is not equal to the total number of protons, giving it a net positive or negative electrical charge. Since protons are positively charged and electrons are negatively charged, if there are more electrons than protons, the atom or molecule will be negatively charged. This is called an an Ion , from the Greek á¼€νÎ¬ , meaning "up".
Solution	In chemistry, a Solution is a homogeneous mixture composed of two or more substances. In such a mixture, a solute is dissolved in another substance, known as a solvent. Gases may dissolve in liquids, for example, carbon dioxide or oxygen in water.
Iron	Iron is a chemical element with the symbol Fe and atomic number 26. Iron is a group 8 and period 4 element. Iron and Iron alloys (steels) are by far the most common metals and the most common ferromagnetic materials in everyday use.
Nitric acid	Nitric acid, also known as aqua fortis and spirit of nitre, is a highly corrosive and toxic strong acid that can cause severe burns. Colorless when pure, older samples tend to acquire a yellow cast due to the accumulation of oxides of nitrogen. If the solution contains more than 86% Nitric acid, it is referred to as fuming Nitric acid.
Acid	An Acid is traditionally considered any chemical compound that, when dissolved in water, gives a solution with a hydrogen ion activity greater than in pure water, i.e. a pH less than 7.0. That approximates the modern definition of Johannes Nicolaus Brønsted and Martin Lowry, who independently defined an Acid as a compound which donates a hydrogen ion (H^+) to another compound (called a base.) Common examples include acetic Acid and sulfuric Acid (used in car batteries.)
Galvanic cell	The Galvanic cell is a part of a battery consisting of an electrochemical cell with two different metals connected by a salt bridge or a porous disk between the individual half-cells. It is sometimes also called a Voltaic cell.

Chapter 19. Electrochemistry

Chapter 19. Electrochemistry

	Common usage of the word battery has evolved to include a single Galvanic cell but the first batteries had many Galvanic cell s.
Salt	Salt is a dietary mineral composed primarily of sodium chloride that is essential for animal life, but toxic to most land plants. Salt flavor is one of the basic tastes, an important preservative and a popular food seasoning.
Salt bridge	A Salt bridge in chemistry, is a laboratory device used to connect the oxidation and reduction half-cells of a galvanic cell (voltaic cell), a type of electrochemical cell. Salt bridge s usually come in two types: glass tube and filter paper. One type of Salt bridge consists of a U-shaped glass tube filled with a relatively inert electrolyte, usually potassium chloride or sodium chloride is used, although the diagram here illustrates the use of a potassium nitrate solution.
Cadmium	Cadmium is a chemical element with the symbol Cd and atomic number 48. The soft, bluish-white transition metal is chemically similar to the two other metals in group 12 , zinc and mercury. Similar to zinc it prefers oxidation state +2 in most of its compounds and similar to mercury it shows a low melting point for a transition metal.
Silver	Silver is a chemical element with the chemical symbol Ag and atomic number 47. A soft, white, lustrous transition metal, it has the highest electrical conductivity of any element and the highest thermal conductivity of any metal. The metal occurs naturally in its pure, free form , as an alloy with gold (electrum) and other metals, and in minerals such as argentite and chlorargyrite.
Atomic mass	The Atomic mass is the mass of an atom, most often expressed in unified Atomic mass units. The Atomic mass may be considered to be the total mass of protons, neutrons and electrons in a single atom (when the atom is motionless.) The Atomic mass is sometimes incorrectly used as a synonym of relative Atomic mass, average Atomic mass and atomic weight; however, these differ subtly from the Atomic mass.
Configuration	The configuration of a molecule is the permanent geometry that results from the spatial arrangement of its bonds. The ability of the same set of atoms to form two or more molecules with different configurations is stereoisomerism. configuration is distinct from chemical conformation, a shape attainable by bond rotations.
Electron	The Electron is a subatomic particle that carries a negative electric charge. It has no known substructure and is believed to be a point particle. An Electron has a mass that is approximately 1836 times less than that of the proton.
Electron configuration	In atomic physics and quantum chemistry, Electron configuration is the arrangement of electrons of an atom, a molecule, or other physical structure. It concerns the way electrons can be distributed in the orbitals of the given system (atomic or molecular for instance.)

Chapter 19. Electrochemistry

Chapter 19. Electrochemistry

Like other elementary particles, the electron is subject to the laws of quantum mechanics, and exhibits both particle-like and wave-like nature.

Oxidation state

In chemistry, the Oxidation state is an indicator of the degree of oxidation of an atom in a chemical compound. The formal Oxidation state is the hypothetical charge that an atom would have if all bonds to atoms of different elements were 100% ionic. Oxidation state s are typically represented by integers, which can be positive, negative, or zero.

Reducing agent

A Reducing agent is the element or compound in a redox reaction that reduces another species. In doing so, it becomes oxidized, and is therefore the electron donor in the redox. For example consider the following reaction:

$[Fe_6]^{4-} + 1/2\ Cl_2 \rightarrow [Fe_6]^{3-} + Cl^-$

The Reducing agent in this reaction is ferrocyanide: it donates an electron, converting to ferricyanide, simultaneous with the reduction of chlorine to chloride.

Spontaneous process

A Spontaneous process is the time-evolution of a system in which it releases free energy (most often as heat) and moves to a lower, more thermodynamically stable, energy state. The sign convention of changes in free energy follows the general convention for thermodynamic measurements, in which a release of free energy from the system corresponds to a negative change in free energy, but a positive change for the surroundings.

A Spontaneous process is capable of proceeding in a given direction, as written or described, without needing to be driven by an outside source of energy.

Concentration

In chemistry, Concentration is the measure of how much of a given substance there is mixed with another substance. This can apply to any sort of chemical mixture, but most frequently the concept is limited to homogeneous solutions, where it refers to the amount of solute in the solvent. To concentrate a solution, one must add more solute, or reduce the amount of solvent (for instance, by selective evaporation.)

Buffer solution

A Buffer solution is an aqueous solution consisting of a mixture of a weak acid and its conjugate base or a weak base and its conjugate acid. It has the property that the pH of the solution changes very little when a small amount of acid or base is added to it. Buffer solution s are used as a means of keeping pH at a nearly constant value in a wide variety of chemical applications.

PH meter

A PH meter is an electronic instrument used to measure the pH (acidity or alkalinity) of a liquid (though special probes are sometimes used to measure the pH of semi-solid substances.) A typical PH meter consists of a special measuring probe (a glass electrode) connected to an electronic meter that measures and displays the pH reading.

The pH probe measures pH as the activity of hydrogen ions surrounding a thin-walled glass bulb at its tip.

Chapter 19. Electrochemistry

Chapter 19. Electrochemistry

Group	In chemistry, a Group is a vertical column in the periodic table of the chemical elements. The name family is derived from the fact that the elements share similar characteristics and traits, just as members of any human family would. There are 18 groups in the standard periodic table.
Lead	Lead is a main-group element with symbol Pb and atomic number 82. Lead is a soft, malleable poor metal, also considered to be one of the heavy metals. Lead has a bluish-white color when freshly cut, but tarnishes to a dull grayish color when exposed to air.
Lithium	Lithium is the chemical element with atomic number 3, and is represented by the symbol Li. It is a soft alkali metal with a silver-white color. Under standard conditions it is the lightest metal and the least dense solid element.
Oxygen	Oxygen and -γενÎ®ς (-genÄ"s) (producer, literally begetter) is the element with atomic number 8 and represented by the symbol O. It is a member of the chalcogen group on the periodic table, and is a highly reactive nonmetallic period 2 element that readily forms compounds (notably oxides) with almost all other elements. At standard temperature and pressure two atoms of the element bind to form di Oxygen , a colorless, odorless, tasteless diatomic gas with the formula O_2. Oxygen is the third most abundant element in the universe by mass after hydrogen and helium and the most abundant element by mass in the Earth"s crust.
Cathodic protection	Cathodic protection is a technique to control the corrosion of a metal surface by making it work as a cathode of an electrochemical cell. This is achieved by placing in contact with the metal to be protected another more easily corroded metal to act as the anode of the electrochemical cell. Cathodic protection systems are most commonly used to protect steel, water or fuel pipelines and storage tanks, steel pier piles, ships, offshore oil platforms and onshore oil well casings.
Pipes	PIPES is the common name for piperazine-N,N'-bis(2-ethanesulfonic acid), a frequently used buffering agent in biochemistry. It is an ethanesulfonic acid buffer developed by Good et al. in the 1960s.
Downs cell	The Downs process is a method for the commercial preparation of metallic sodium, in which molten NaCl is electrolyzed in a special apparatus called the Downs cell. Schematic diagram of the Downs cell The Downs cell uses a carbon anode and iron cathode. The electrolyte is sodium chloride that has been fused to a liquid by heating.
Sodium	Sodium is a metallic element with a symbol Na and atomic number 11. It is a soft, silvery-white, highly reactive metal and is a member of the alkali metals within "group 1" (formerly known as "group IA".) It has only one stable isotope, ^{23}Na.

Chapter 19. Electrochemistry

Chapter 19. Electrochemistry

Metallurgy	Metallurgy is a domain of materials science that studies the physical and chemical behavior of metallic elements, their intermetallic compounds, and their mixtures, which are called alloys. It is also the technology of metals: the way in which science is applied to their practical use. Metallurgy is commonly used in the craft of metalworking.
Hydroxide	In chemistry, Hydroxide is the name for the diatomic anion OH⁻, consisting of oxygen and hydrogen atoms, usually derived from the dissociation of a base. It is one of the simplest diatomic ions known. Inorganic compounds that contain the hydroxyl group are referred to as Hydroxide s.
Sodium hydroxide	Sodium hydroxide is a caustic metallic base. Sodium hydroxide forms a strong alkaline solution when dissolved in a solvent such as water. However, only the hydroxide ion is basic.
Water	Water is the most abundant molecule on Earth"s surface, constituting about 75% of the Earth"s surface. In nature it exists in liquid, solid, and gaseous states. It is in dynamic equilibrium between the liquid and gas states at standard temperature and pressure.
Acid rain	Acid rain is rain or any other form of precipitation that is unusually acidic. It has harmful effects on plants, aquatic animals, and infrastructure. Acid rain is mostly caused by human emissions of sulfur and nitrogen compounds which react in the atmosphere to produce acids.
Sulfuric acid	Sulfuric (or sulphuric) acid, H_2SO_4, is a strong mineral acid. It is soluble in water at all concentrations. Sulfuric acid has many applications, and is one of the top products of the chemical industry.
Electrolysis of Water	Electrolysis of water is the decomposition of water (H_2O) into oxygen (O_2) and hydrogen gas (H_2) due to an electric current being passed through the water. This electrolytic process is rarely used in industrial applications since hydrogen can be produced more affordably from fossil fuels. An electrical power source is connected to two electrodes, or two plates, which are placed in the water.
Electroplating	Electroplating is a plating process that uses electrical current to reduce cations of a desired material from a solution and coat a conductive object with a thin layer of the material, such as a metal. Electroplating is primarily used for depositing a layer of material to bestow a desired property (e.g., abrasion and wear resistance, corrosion protection, lubricity, aesthetic qualities, etc.) to a surface that otherwise lacks that property.
Metal	In chemistry, a Metal is an element, compound, or alloy characterized by high electrical conductivity. In a Metal atoms readily lose electrons to form positive ions ; those ions are surrounded by delocalized electrons, which are responsible for the conductivity. The thus produced solid is held by electrostatic interactions between the ions and the electron cloud, which are called Metal lic bonds.

Chapter 19. Electrochemistry

Chapter 19. Electrochemistry

Mercury	There are seven isotopes of mercury with Hg-202 being the most abundant (29.86%.) The longest-lived radioisotopes are ^{194}Hg with a half-life of 444 years, and ^{203}Hg with a half-life of 46.612 days. Most of the remaining radioisotopes have half-lives that are less than a day.
Stoichiometry	Stoichiometry is the calculation of quantitative (measurable) relationships of the reactants and products in a balanced chemical reaction (chemicals.) It can be used to calculate quantities such as the amount of products that can be produced with the given reactants and percent yield. "Stoichiometry" is derived from the Greek words στοιχεά¿–ον and μÎ¼τρον (metron, meaning measure.)
Potassium	Potassium is a chemical element. It has the symbol K , atomic number 19, and atomic mass 39.0983. Potassium was first isolated from potash.

Chapter 19. Electrochemistry

Chapter 20. Nuclear Chemistry

Becquerel	The Becquerel is the SI derived unit of radioactivity. One Bq is defined as the activity of a quantity of radioactive material in which one nucleus decays per second. It is therefore equivalent to s^{-1}.
Molybdenum	Molybdenum, is a Group 6 chemical element with the symbol Mo and atomic number 42. The free element, which is a silvery metal, has the sixth-highest melting point of any element. It readily forms hard, stable carbides, and for this reason it is often used in high-strength steel alloys.
Nuclear chemistry	Nuclear chemistry is a subfield of chemistry dealing with radioactivity, nuclear processes and nuclear properties. • It is the chemistry of radioactive elements such as the actinides, radium and radon together with the chemistry associated with equipment (such as nuclear reactors) which are designed to perform nuclear processes. This includes the corrosion of surfaces and the behaviour under conditions of both normal and abnormal operation (such as during an accident.) An important area is the behaviour of objects and materials after being placed into a waste store or otherwise disposed of. • the study of the chemical effects resulting from the absorption of radiation within living animals, plants, and other materials. The radiation chemistry controls much of radiation biology as radiation has an effect on living things at the molecular scale, to explain it another way the radiation alters the biochemicals within an organism, the alteration of the biomolecules then changes the chemistry which occurs within the organism, this change in biochemistry then can lead to a biological outcome. As a result Nuclear chemistry greatly assists the understanding of medical treatments (such as cancer radiotherapy) and has enabled these treatments to improve.
Radioactive decay	Radioactive decay is the process in which an unstable atomic nucleus spontaneously loses energy by emitting ionizing particles and radiation. This decay, or loss of energy, results in an atom of one type, called the parent nuclide transforming to an atom of a different type, called the daughter nuclide. For example: a carbon-14 atom (the "parent") emits radiation and transforms to a nitrogen-14 atom (the "daughter".)
Radioactivity	Radioactivity can be used in life sciences as a radiolabel to easily visualise components or target molecules in a biological system. Radionuclei are synthesised in particle accelerators and have short half-lives, giving them high maximum theoretical specific activities. This lowers the detection time compared to radionuclei with longer half-lives, such as carbon-14.
Technetium	Technetium is the lightest chemical element with no stable isotope, and therefore the lightest radioactive element. It is a synthetic element with the atomic number 43 and is given the symbol Tc. The chemical properties of this silvery grey, crystalline transition metal are intermediate between rhenium and manganese.

Chapter 20. Nuclear Chemistry

Chapter 20. Nuclear Chemistry

Chemistry	In the history of science, the etymology of the word Chemistry is a debatable issue. It is agreed that the word "alchemy" is a European one, derived from Arabic, but the origin of the root word, chem, is uncertain. Words similar to it have been found in most ancient languages, with different meanings, but conceivably somehow related to alchemy.
Isotopes	Isotopes are any of the different types of atoms of the same chemical element, each having a different atomic mass (mass number.) Isotopes of an element have nuclei with the same number of protons (the same atomic number) but different numbers of neutrons. Therefore, Isotopes of the same element have different mass numbers (number of nucleons.)
Nuclear reaction	In nuclear physics, a Nuclear reaction is the process in which two nuclei or nuclear particles collide to produce products different from the initial particles. In principle a reaction can involve more than three particles colliding, but because the probability of three or more nuclei to meet at the same time at the same place is much less than for two nuclei, such an event is exceptionally rare. While the transformation is spontaneous in the case of radioactive decay, it is initiated by a particle in the case of a Nuclear reaction.
Nuclide	A Nuclide is an atomic nucleus characterized by its specific constitution, i.e., by its number of protons, its number of neutrons, and its energy content. The various Nuclide s, or species, of a particular chemical element with equal proton number , but different neutron numbers were called isotopes of the element, before the more inclusive term Nuclide was internationally accepted (ca. 1950.)
Uranium	Uranium is a silvery-white metallic chemical element in the actinide series of the periodic table that has the symbol U and atomic number 92. Besides its 92 protons, a Uranium nucleus can have between 141 and 146 neutrons, with 146 and 143 (U-235) in its most common isotopes. The number of electrons in a Uranium atom is 92, 6 of them valence electrons.
Background radiation	Background radiation is constantly present in the environment and is emitted from a variety of natural and artificial sources. Primary contributions come from: - Sources in the earth. These include sources in food and water, which are incorporated in the body, and in building materials and other products that incorporate those radioactive sources; - Sources from space, in the form of cosmic rays; - Sources in the atmosphere. One significant contribution comes from the radon gas that is released from the Earth"s crust and subsequently decays into radioactive atoms that become attached to airborne dust and particulates. Another contribution arises from the radioactive atoms produced in the bombardment of atoms in the upper atmosphere by high-energy cosmic rays.

Chapter 20. Nuclear Chemistry

Chapter 20. Nuclear Chemistry

About 3% of Background radiation comes from other man-made sources such as:

- Self-luminous dials and signs
- Global radioactive contamination due to historical nuclear weapons testing
- Nuclear power station or nuclear fuel reprocessing accidents (though these are rare)
- Normal operation of facilities used for nuclear power and scientific research
- Emissions from burning fossil fuels, such as coal fired power plants
- Emissions from nuclear medicine facilities and patients
- Emissions from the improper disposal or recycling of radioactive materials used in nuclear medicine

Accidental exposure to man-made radioactive substances can result in radiation exposure that is many times that received from background sources, whether natural or man-made. Additionally, radiation therapy can cause relatively high levels of exposure. However, when it comes to Background radiation, naturally occurring sources are responsible for the vast majority of radiation exposure.

Emission

In physics, Emission is the process by which the energy of a photon is released by another entity, for example, by an atom whose electrons make a transition between two electronic energy levels. The emitted energy is in the form of a photon.

The emittance of an object quantifies how much light is emitted by it.

Nucleon

In physics, a Nucleon is a collective name for two baryons: the neutron and the proton. They are constituents of the atomic nucleus and until the 1960s were thought to be elementary particles. In those days their interactions (now called inter Nucleon interactions) defined strong interactions.

Radium

Radium is a radioactive chemical element which has the symbol Ra and atomic number 88. Its appearance is almost pure white, but it readily oxidizes on exposure to air, turning black. Radium is an alkaline earth metal that is found in trace amounts in uranium ores.

Product

A Product is a substance that forms as a result of a biological- or chemical reaction. While the end Product of some chemical reactions may be the result of a relatively rapid reaction, nanoseconds to seconds, chemical equilibria in complex systems may require years or even centuries to be established. For example, equilibria in groundwater systems with multiple components are achieved on timescales of millennia, if ever.

Magic number

In nuclear physics, a Magic number is a number of nucleons (either protons or neutrons) such that they are arranged into complete shells within the atomic nucleus. The seven most widely recognised magic numbers as of 2007 are

2, 8, 20, 28, 50, 82, 126. (sequence A018226 in OEIS)

Chapter 20. Nuclear Chemistry

Chapter 20. Nuclear Chemistry

	Atomic nuclei consisting of such a Magic number of nucleons have a higher average binding energy per nucleon than one would expect based upon predictions such as the semi-empirical mass formula and are hence more stable against nuclear decay.
Nuclear force	The Nuclear force is the force between two or more nucleons. It is responsible for binding of protons and neutrons into atomic nuclei. To a large extent, this force can be understood in terms of the exchange of virtual light mesons, such as the pions.
Band of stability	The Band of stability is the range of the number of neutrons against the number of protons for stable nuclei graph that plots all stable nuclei. The Band of stability can be used when looking at isotopes in order to determine how a chemical element will undergo radioactive decay. In order to determine where an element lands on the graph, one must look at the periodic table of elements.
Polonium	Polonium is a chemical element with the symbol Po and atomic number 84, discovered in 1898 by Marie and Pierre Curie. A rare and highly radioactive metalloid, Polonium is chemically similar to bismuth and tellurium, and it occurs in uranium ores. Polonium has been studied for possible use in heating spacecraft.
Carbon	Carbon is the chemical element with symbol C and atomic number 6. As a member of group 14 on the periodic table, it is nonmetallic and tetravalent--making four electrons available to form covalent chemical bonds. There are three naturally occurring isotopes, with ^{12}C and ^{13}C being stable, while ^{14}C is radioactive, decaying with a half-life of about 5730 years.
Electron	The Electron is a subatomic particle that carries a negative electric charge. It has no known substructure and is believed to be a point particle. An Electron has a mass that is approximately 1836 times less than that of the proton.
Electron capture	Electron capture is a decay mode for isotopes that will occur when there are too many protons in the nucleus of an atom and insufficient energy to emit a positron; however, it continues to be a viable decay mode for radioactive isotopes that can decay by positron emission. If the energy difference between the parent atom and the daughter atom is less than 1.022 MeV, positron emission is forbidden and Electron capture is the sole decay mode. For example, Rubidium-83 will decay to Krypton-83 solely by Electron capture (the energy difference is about 0.9 MeV.)
Positron emission	Positron emission is a type of beta decay, sometimes referred to as "beta plus" (β^+.) In beta plus decay, a proton is converted, via the weak force, to a neutron, a positron (also known as the "beta plus particle", the antimatter counterpart of an electron), and a neutrino. Isotopes which undergo this decay and thereby emit positrons include carbon-11, potassium-40, nitrogen-13, oxygen-15, fluorine-18, and iodine-121.
Potassium	Potassium is a chemical element. It has the symbol K, atomic number 19, and atomic mass 39.0983. Potassium was first isolated from potash.

Chapter 20. Nuclear Chemistry

Chapter 20. Nuclear Chemistry

Acid	An Acid is traditionally considered any chemical compound that, when dissolved in water, gives a solution with a hydrogen ion activity greater than in pure water, i.e. a pH less than 7.0. That approximates the modern definition of Johannes Nicolaus Brønsted and Martin Lowry, who independently defined an Acid as a compound which donates a hydrogen ion (H^+) to another compound (called a base.) Common examples include acetic Acid and sulfuric Acid (used in car batteries.)
Nuclear fission	In nuclear physics and nuclear chemistry, Nuclear fission is a nuclear reaction in which the nucleus of an atom splits into smaller parts, often producing free neutrons and lighter nuclei, which may eventually produce photons (in the form of gamma rays.) Fission of heavy elements is an exothermic reaction which can release large amounts of energy both as electromagnetic radiation and as kinetic energy of the fragments (heating the bulk material where fission takes place.) For fission to produce energy, the total binding energy of the resulting elements has to be higher than that of the starting element.
Phosphorous acid	Phosphorous acid is the compound described by the formula H_3PO_3. It can be formulated as HP(O)(OH)$_2$ and therefore contains phosphorus in oxidation state +3. It is one of the oxoacids of phosphorus, other important members being phosphoric acid (H_3PO_4) and hypo Phosphorous acid (H_3PO_2.)
Phosphorus	Phosphorus is the chemical element that has the symbol P and atomic number 15. A multivalent nonmetal of the nitrogen group, Phosphorus is commonly found in inorganic phosphate rocks. Elemental Phosphorus exists in two major forms - white Phosphorus and red Phosphorus.
Spontaneous fission	Spontaneous fission is a form of radioactive decay characteristic of very heavy isotopes. It is theoretically possible for any atomic nucleus whose mass is greater than or equal to 100 atomic mass units (u), i.e. elements near ruthenium. In practice, however, Spontaneous fission is only energetically feasible for atomic masses above 230 u (elements near thorium.)
Ionization	Ionization is the physical process of converting an atom or molecule into an ion by adding or removing charged particles such as electrons or other ions. This is often confused with dissociation (chemistry.) The process works slightly differently depending on whether an ion with a positive or a negative electric charge is being produced.
Beryllium	Beryllium is the chemical element with the symbol Be and atomic number 4. A bivalent element, Beryllium is found naturally only combined with other elements in minerals. Notable gemstones which contain Beryllium include Beryl (aquamarines and emeralds) and Chrysoberyl (Alexandrite and Cat"s eye.)

Chapter 20. Nuclear Chemistry

Chapter 20. Nuclear Chemistry

Rutherford	The Rutherford is an obsolete unit of radioactivity, defined as the activity of a quantity of radioactive material in which one million nuclei decay per second. It is therefore equivalent to one megabecquerel. It was named after Ernest Rutherford It is not an SI unit.
Particle	In marine and freshwater ecology, a Particle is a small object. Particles can remain in suspension in the ocean or freshwater, however they eventually settle (rate determined by Stokes" law) and accumulate as sediment. Some can enter the atmosphere through wave action where they can act as cloud condensation nuclei (CCN.)
Neptunium	Neptunium is a chemical element with the symbol Np and atomic number 93. A radioactive metallic element, Neptunium is the first transuranic element and belongs to the actinide series. Its most stable isotope, ^{237}Np, is a by-product of nuclear reactors and plutonium production and it can be used as a component in neutron detection equipment.
Plutonium	Plutonium is a rare transuranic radioactive element. It is an actinide metal of silvery-white appearance that tarnishes when exposed to air, forming a dull coating when oxidized. The element normally exhibits six allotropes and four oxidation states.
Transuranium elements	In chemistry, Transuranium elements are the chemical elements with atomic numbers greater than 92 None of these elements are stable; they decay radioactively into other elements. Of the elements with atomic numbers 1 to 92, all but four occur in easily detectable quantities on earth, having stable, or very long half life isotopes, or are created as common products of the decay of uranium.
Activity	In chemical thermodynamics Activity is a measure of the "effective concentration" of a species in a mixture. By convention, it is a dimensionless quantity. The Activity of pure substances in condensed phases (solid or liquids) is normally taken as unity.
Actinium	Actinium is a radioactive chemical element with the symbol Ac and atomic number 89, which was discovered in 1899. It was the first non-primordial radioactive element to be isolated, although polonium, radium and radon were observed before, but not isolated until 1902. It gave the name to the actinoid series, a group of 15 similar elements between Actinium and lawrencium in the periodic table.
Americium	Americium is a synthetic element that has the symbol Am and atomic number 95. A radioactive metallic element, Americium is an actinide that was obtained in 1944 by Glenn T. Seaborg who was bombarding plutonium with neutrons and was the fourth transuranic element to be discovered. It was named for the Americas, by analogy with europium.
Curium	Curium is a synthetic chemical element with the symbol Cm and atomic number 96. A radioactive metallic transuranic element of the actinide series, Curium is produced by bombarding plutonium with alpha particles and was named for Marie Curie and her husband Pierre.

Chapter 20. Nuclear Chemistry

Chapter 20. Nuclear Chemistry

	The isotope Curium-248 has been synthesized only in milligram quantities, but Curium-242 and Curium-244 are made in multigram amounts, which allows for the determination of some of the element"s properties.
Albert Ghiorso	Albert Ghiorso is an American nuclear scientist who helped discover numerous chemical elements on the periodic table.
	He was born in Vallejo, California and grew up in Alameda, California. As a teenager, he built radio circuitry and earned a reputation for establishing radio contacts at distances that outdid the military.
Glenn Theodore Seaborg	Glenn Theodore Seaborg was an American scientist who won the 1951 Nobel Prize in Chemistry for "discoveries in the chemistry of the transuranium elements," contributed to the discovery and isolation of ten elements, developed the actinoids concept and was the first to propose the actinoids series which led to the current arrangement of the Periodic Table of the Elements. He spent most of his career as an educator and research scientist at the University of California, Berkeley where he became the second Chancellor in its history and served as a University Professor. Seaborg advised ten presidents from Truman to Clinton on nuclear policy and was the chairman of the United States Atomic Energy Commission from 1961 to 1971 where he pushed for commercial nuclear energy and peaceful applications of nuclear science.
Phosphor	A Phosphor is a substance that exhibits the phenomenon of Phosphor escence (sustained glowing after exposure to energized particles such as electrons or ultraviolet photons.)
	Phosphor s are transition metal compounds or rare earth compounds of various types. The most common uses of Phosphor s are in CRT displays and fluorescent lights.
Curie	The Curie is a unit of radioactivity, defined as
	$\qquad$ 1 Ci = 3.7×10^{10} decays per second or becquerels.
	This is roughly the activity of 1 gram of the radium isotope ^{226}Ra, a substance studied by the pioneers of radiology, Marie and Pierre Curie. The Curie has since been replaced by an SI derived unit, the becquerel (Bq), which equates to one decay per second.
Polymer	A Polymer is a large molecule composed of repeating structural units typically connected by covalent chemical bonds. While Polymer in popular usage suggests plastic, the term actually refers to a large class of natural and synthetic materials with a variety of properties.
	Due to the extraordinary range of properties accessible in Polymer ic materials , they have come to play an essential and ubiquitous role in everyday life - from plastics and elastomers on the one hand to natural bio Polymer s such as DNA and proteins that are essential for life on the other.
Source	Source Holdings Ltd. is a specialist European-based provider of Exchange Traded Products (ETPs), including Exchange Traded Funds (ETFs) and Exchange Traded Commodities (ETCs.) The company was founded in 2008 by Bank of America Merrill Lynch, Goldman Sachs and Morgan Stanley and is headquartered inside the City of London, on 88 Wood Street.

Chapter 20. Nuclear Chemistry

Chapter 20. Nuclear Chemistry

	Source launched its first product offering on the 20th April 2009, although the 3 founders are reported to have started working on the project in early 2008.
Half-life	The Half-life of a quantity whose value decreases with time is the interval required for the quantity to decay to half of its initial value. The concept originated in describing how long it takes atoms to undergo radioactive decay but also applies in a wide variety of other situations. The term "Half-life" dates to 1907.
Tritium	Tritium is a radioactive isotope of hydrogen. The nucleus of Tritium contains one proton and two neutrons, whereas the nucleus of protium (the most abundant hydrogen isotope) contains one proton and no neutrons. While Tritium has several different experimentally-determined values of its half-life, the NIST recommends 4,500±8 days (approximately 12.33 years.)
Melvin Ellis Calvin	Melvin Ellis Calvin was an American chemist most famed for discovering the Calvin cycle along with Andrew Benson and James Bassham, for which he was awarded the 1961 Nobel Prize in Chemistry. He spent most of his five-decade career at the University of California, Berkeley. Calvin was born in St. Paul, Minnesota, the son of Jewish immigrants.
Photosynthesis	Photosynthesis is a process that converts carbon dioxide into organic compounds, especially sugars, using the energy from sunlight. Photosynthesis occurs in plants, algae, and many species of Bacteria, but not in Archaea. Photosynthetic organisms are called photoautotrophs, since it allows them to create their own food.
Solution	In chemistry, a Solution is a homogeneous mixture composed of two or more substances. In such a mixture, a solute is dissolved in another substance, known as a solvent. Gases may dissolve in liquids, for example, carbon dioxide or oxygen in water.
Arsenic	Arsenic is the chemical element that has the symbol As and atomic number 33. Arsenic was first documented by Albertus Magnus in 1250. Its atomic mass is 74.92.
Cobalt	Cobalt is a hard, lustrous, grey metal, a chemical element with symbol Co and atomic number 27. Although Cobalt based colors and pigments have been used since ancient times for making jewelry and paints, and miners have long used the name kobold ore for some minerals, the free metallic Cobalt was not prepared and discovered until 1735 by Georg Brandt.
Catalyst	Catalyst is a science centre and museum devoted to the chemical industry. Its full title is Catalyst Science Discovery Centre. It is located in Widnes, Cheshire, in the north west of England, and situated on the north bank of the River Mersey (grid reference SJ512841.)

Chapter 20. Nuclear Chemistry

Chapter 20. Nuclear Chemistry

Buffer solution

A Buffer solution is an aqueous solution consisting of a mixture of a weak acid and its conjugate base or a weak base and its conjugate acid. It has the property that the pH of the solution changes very little when a small amount of acid or base is added to it. Buffer solution s are used as a means of keeping pH at a nearly constant value in a wide variety of chemical applications.

Resonance

Resonance in chemistry is a key component of valence bond theory used to graphically represent and mathematically model certain types of molecular structures when no single, conventional Lewis structure can satisfactorily represent the observed structure or explain its properties. Resonance instead considers such molecules to be an intermediate or average (called a Resonance hybrid) between several Lewis structures that differ only in the placement of the valence electrons. Scheme 1.

Einstein

An Einstein is a unit used in irradiance and in photochemistry. One Einstein is defined as one mole of photons, regardless of their frequency. Therefore, the number of photons in an Einstein is Avogadro"s number.

Binding

Molecular Binding is an attractive interaction between two molecules which results in a stable association in which the molecules are in close proximity to each other. The result of molecular Binding is formation of a molecular complex.
Molecular Binding can be classified into the following types:

- non-covalent - no chemical bonds are formed between the two interacting molecules hence the association is fully reversible
- reversible covalent - a chemical bond is formed, however the free energy difference separating the non-bonded reactants from bonded product is near equilibrium and the activation barrier is relatively low such that the reverse reaction which cleaves the chemical bond easily occurs
- irreversible covalent - a chemical bond is formed in which the product is thermodynamically much more stable than the reactants such that the reverse reaction does not take place

Molecules that can participate in molecular Binding include proteins, nucleic acids, carbohydrates, lipids, and small organic molecules such as drugs. Hence the types of complexes that form as a result of molecular Binding include:

- protein - protein
- protein - DNA
- protein - hormone
- protein - drug

Proteins that form stable complexes with other molecules are often referred to as receptors while their Binding partners are called ligands.

Chapter 20. Nuclear Chemistry

Chapter 20. Nuclear Chemistry

Binding energy	Binding energy is the mechanical energy required to disassemble a whole into separate parts. A bound system has typically a lower potential energy than its constituent parts; this is what keeps the system together. The usual convention is that this corresponds to a positive Binding energy.
Californium	Californium is a metallic chemical element with the symbol Cf and atomic number 98. A radioactive transuranic element, Californium is used in starting nuclear reactors, optimizing coal-fired power plants and cement production facilities, medical treatment of cancer, and oil exploration via down hole well logging. It was first produced by bombarding curium with alpha particles (helium ions.)
Otto Hahn	Otto Hahn was a German chemist and Nobel laureate who pioneered the fields of radioactivity and radiochemistry. He is regarded as "the father of nuclear chemistry" and the "founder of the atomic age".
	Hahn was the youngest son of Heinrich Hahn, a prosperous glazier and entrepreneur ("Glasbau Hahn"), and Charlotte Hahn, née Giese (1845-1905.)
Nuclear fusion	In nuclear physics and nuclear chemistry, Nuclear fusion is the process by which multiple like-charged atomic nuclei join together to form a heavier nucleus. It is accompanied by the release or absorption of energy, which allows matter to enter a plasma state.
	The fusion of two nuclei with lower mass than iron (which, along with nickel, has the largest binding energy per nucleon) generally releases energy while the fusion of nuclei heavier than iron absorbs energy; vice-versa for the reverse process, nuclear fission.
Critical mass	A Critical mass is the smallest amount of fissile material needed for a sustained nuclear chain reaction. The Critical mass of a fissionable material depends upon its nuclear properties (e.g. the nuclear fission cross-section), its density, its shape, its enrichment, its purity, its temperature, and its surroundings.
	The term critical refers to an equilibrium fission reaction (steady-state or continuous chain reaction); this is where there is no increase or decrease in power, temperature, or neutron population.
Deuterium	Deuterium, also called heavy hydrogen, is a stable isotope of hydrogen with a natural abundance in the oceans of Earth of approximately one atom in 6,500 of hydrogen (~154 ppm.) Deuterium thus accounts for approximately 0.015% (alternately, on a weight basis: 0.031%) of all naturally occurring hydrogen in the oceans on Earth Deuterium abundance on Jupiter is about 2.25×10^{-5} (roughly 22 atoms in a million, or 15% of the terrestrial Deuterium-to-hydrogen ratio); these ratios presumably reflect the early solar nebula ratios, and those after the Big Bang.

Chapter 20. Nuclear Chemistry

Chapter 20. Nuclear Chemistry

Hydrogen | Hydrogen is the chemical element with atomic number 1. It is represented by the symbol H. At standard temperature and pressure, Hydrogen is a colorless, odorless, nonmetallic, tasteless, highly flammable diatomic gas with the molecular formula H_2. With an atomic weight of 1.007 94 u, Hydrogen is the lightest element.

Chapter 20. Nuclear Chemistry

Chapter 21. Chemistry of the Main-Group Elements

Carbon	Carbon is the chemical element with symbol C and atomic number 6. As a member of group 14 on the periodic table, it is nonmetallic and tetravalent--making four electrons available to form covalent chemical bonds. There are three naturally occurring isotopes, with ^{12}C and ^{13}C being stable, while ^{14}C is radioactive, decaying with a half-life of about 5730 years.
Hydrogen	Hydrogen is the chemical element with atomic number 1. It is represented by the symbol H. At standard temperature and pressure, Hydrogen is a colorless, odorless, nonmetallic, tasteless, highly flammable diatomic gas with the molecular formula H_2. With an atomic weight of 1.007 94 u, Hydrogen is the lightest element.
Oxygen	Oxygen and -γενÎ®ς (-genÄ"s) (producer, literally begetter) is the element with atomic number 8 and represented by the symbol O. It is a member of the chalcogen group on the periodic table, and is a highly reactive nonmetallic period 2 element that readily forms compounds (notably oxides) with almost all other elements. At standard temperature and pressure two atoms of the element bind to form di Oxygen , a colorless, odorless, tasteless diatomic gas with the formula O_2. Oxygen is the third most abundant element in the universe by mass after hydrogen and helium and the most abundant element by mass in the Earth"s crust.
Periodic trends	In Chemistry, Periodic trends are the tendencies of certain elemental characteristics to increase or decrease as one progresses from one corner of the Periodic table of elements. The atomic radius is the distance from the atomic nucleus to the outermost stable electron orbital in an atom that is at equilibrium. The atomic radius tends to decrease as one progresses across a period because the effective nuclear charge increases, thereby attracting the orbiting electrons and lessening the radius.
Abundance	The abundance of a chemical element measures how relatively common the element is, or how much of the element there is by comparison to all other elements. abundance may be variously measured by the mass-fraction (the same as weight fraction), or mole-fraction (fraction of atoms, or sometimes fraction of molecules, in gases), or by volume fraction. Measurement by volume-fraction is a common abundance measure in mixed gases such as atmospheres, which is close to molecular mole-fraction for ideal gas mixtures (i.e., gas mixtures at relatively low densities and pressures.)
Acid	An Acid is traditionally considered any chemical compound that, when dissolved in water, gives a solution with a hydrogen ion activity greater than in pure water, i.e. a pH less than 7.0. That approximates the modern definition of Johannes Nicolaus Brønsted and Martin Lowry, who independently defined an Acid as a compound which donates a hydrogen ion (H^+) to another compound (called a base.) Common examples include acetic Acid and sulfuric Acid (used in car batteries.)

Chapter 21. Chemistry of the Main-Group Elements

Chapter 21. Chemistry of the Main-Group Elements

Enthalpy	In thermodynamics and molecular chemistry, the Enthalpy is a thermodynamic property of a fluid. It can be used to calculate the heat transfer during a quasistatic process taking place in a closed thermodynamic system under constant pressure. Enthalpy H is an arbitrary concept but the Enthalpy change ΔH is more useful because it is equal to the change in the internal energy of the system, plus the work that the system has done on its surroundings.
Calcium	Calcium is the chemical element with the symbol Ca and atomic number 20. It has an atomic mass of 40.078 amu. Calcium is a soft grey alkaline earth metal, and is the fifth most abundant element by mass in the Earth"s crust.
Calcium oxide	Calcium oxide, commonly known as burnt lime, lime or quicklime, is a widely used chemical compound. It is a white, caustic and alkaline crystalline solid at room temperature. As a commercial product, lime often also contains magnesium oxide, silicon oxide and smaller amounts of aluminium oxide and iron oxide.
Chromium	Chromium is a chemical element which has the symbol Cr and atomic number 24. It is a steely-gray, lustrous, hard metal that takes a high polish and has a high melting point. It is also odourless, tasteless, and malleable.
Oxidation number	The Oxidation number of a central atom in a coordination compound is the charge that it would have if all the ligands were removed along with the electron pairs that were shared with the central atom. It is used in the nomenclature of inorganic compounds. It is represented by a Roman numeral; the plus sign is omitted for positive Oxidation number s.
Oxidation state	In chemistry, the Oxidation state is an indicator of the degree of oxidation of an atom in a chemical compound. The formal Oxidation state is the hypothetical charge that an atom would have if all bonds to atoms of different elements were 100% ionic. Oxidation state s are typically represented by integers, which can be positive, negative, or zero.
Silicon	Silicon is the most common metalloid. It is a chemical element, which has the symbol Si and atomic number 14. The atomic mass is 28.0855.
Sulfur	Sulfur or sulphur is the chemical element that has the atomic number 16. It is denoted with the symbol S. It is an abundant, multivalent non-metal. Sulfur in its native form is a yellow crystalline solid.
Water	Water is the most abundant molecule on Earth"s surface, constituting about 75% of the Earth"s surface. In nature it exists in liquid, solid, and gaseous states. It is in dynamic equilibrium between the liquid and gas states at standard temperature and pressure.
Magnesium	Magnesium is a chemical element with the symbol Mg, atomic number 12, atomic weight 24.3050 and common oxidation number +2.

Chapter 21. Chemistry of the Main-Group Elements

Chapter 21. Chemistry of the Main-Group Elements

	Magnesium, an alkaline earth metal, is the ninth most abundant element in the universe by mass. The commonness of Magnesium is related to the fact that it is easily built up in supernova stars from a sequential addition of three helium nuclei to carbon .
Sodium	Sodium is a metallic element with a symbol Na and atomic number 11. It is a soft, silvery-white, highly reactive metal and is a member of the alkali metals within "group 1" (formerly known as "group IA".) It has only one stable isotope, ^{23}Na.
Copper	Copper is a chemical element with the symbol Cu and atomic number 29. It is a ductile metal with very high thermal and electrical conductivity. Pure Copper is rather soft and malleable and a freshly-exposed surface has a pinkish or peachy color.
Gibbs free energy	In thermodynamics, the Gibbs free energy is a thermodynamic potential that measures the "useful" or process-initiating work obtainable from an isothermal, isobaric thermodynamic system. The Gibbs free energy is the maximum amount of non-expansion work that can be extracted from a closed system; this maximum can be attained only in a completely reversible process. When a system changes from a well-defined initial state to a well-defined final state, the Gibbs free energy ΔG equals the work exchanged by the system with its surroundings, less the work of the pressure forces, during a reversible transformation of the system from the same initial state to the same final state.
Group	In chemistry, a Group is a vertical column in the periodic table of the chemical elements. The name family is derived from the fact that the elements share similar characteristics and traits, just as members of any human family would. There are 18 groups in the standard periodic table.
Metal	In chemistry, a Metal is an element, compound, or alloy characterized by high electrical conductivity. In a Metal atoms readily lose electrons to form positive ions ; those ions are surrounded by delocalized electrons, which are responsible for the conductivity. The thus produced solid is held by electrostatic interactions between the ions and the electron cloud, which are called Metal lic bonds.
Metallurgy	Metallurgy is a domain of materials science that studies the physical and chemical behavior of metallic elements, their intermetallic compounds, and their mixtures, which are called alloys. It is also the technology of metals: the way in which science is applied to their practical use. Metallurgy is commonly used in the craft of metalworking.
Mineral	A Mineral is a naturally occurring solid formed through geological processes that has a characteristic chemical composition, a highly ordered atomic structure, and specific physical properties. A rock, by comparison, is an aggregate of Mineral s and/or Mineral oids, and need not have a specific chemical composition. Mineral s range in composition from pure elements and simple salts to very complex silicates with thousands of known forms.

Chapter 21. Chemistry of the Main-Group Elements

Chapter 21. Chemistry of the Main-Group Elements

Graphite	Graphite " href="/wiki/Cleavage_(crystal)">loose interlamellar coupling between sheets in the structure, in fact in a vacuum environment (such as in technologies for use in space), Graphite was found to be a very poor lubricant. This fact led to the discovery that Graphite"s lubricity is due to adsorbed air and water between the layers, unlike other layered dry lubricants such as molybdenum disulfide. Recent studies suggest that an effect called superlubricity can also account for this effect.
Hydroxide	In chemistry, Hydroxide is the name for the diatomic anion OH^-, consisting of oxygen and hydrogen atoms, usually derived from the dissociation of a base. It is one of the simplest diatomic ions known. Inorganic compounds that contain the hydroxyl group are referred to as Hydroxide s.
Source	Source Holdings Ltd. is a specialist European-based provider of Exchange Traded Products (ETPs), including Exchange Traded Funds (ETFs) and Exchange Traded Commodities (ETCs.) The company was founded in 2008 by Bank of America Merrill Lynch, Goldman Sachs and Morgan Stanley and is headquartered inside the City of London, on 88 Wood Street. Source launched its first product offering on the 20th April 2009, although the 3 founders are reported to have started working on the project in early 2008 .
Bayer process	The Bayer process is the principal industrial means of refining bauxite to produce alumina. Bauxite, the most important ore of aluminium, contains only 30-54% alumina, Al_2O_3, the rest being a mixture of silica, various iron oxides, and titanium dioxide. The alumina must be purified before it can be refined to aluminium metal.
Gallium	Gallium is a chemical element that has the symbol Ga and atomic number 31. Elemental Gallium does not occur in nature, but as the Ga salt, in trace amounts in bauxite and zinc ores. A soft silvery metallic poor metal, elemental Gallium is a brittle solid at low temperatures.
Gallium arsenide	Gallium arsenide is a compound of two elements, gallium and arsenic. It is an important semiconductor and is used to make devices such as microwave frequency integrated circuits (ie, MMICs), infrared light-emitting diodes, laser diodes and solar cells. Gallium arsenide can be prepared from the elements and a number of industrial processes use this, for example: - the crystal growth using a horizontal zone furnace (Bridgman-Stockbarger technique) where Ga and Arsenic vapor react and deposit on a seed crystal at the cooler end of the furnace. - LEC (liquid encapsulated Czochralski) growth

Chapter 21. Chemistry of the Main-Group Elements

Chapter 21. Chemistry of the Main-Group Elements

Alternative methods for producing films of GaAs include:

- VPE reaction of gaseous gallium metal and arsenic trichloride

$$2Ga + 2AsCl_3 \rightarrow 2GaAs + 3Cl_2$$

- MOCVD reaction of trimethylgallium and arsine:

$$Ga(CH_3)_3 + AsH_3 \rightarrow GaAs + 3CH_4$$

Wet etching of GaAs industrially uses an oxidizing agent e.g. hydrogen peroxide or bromine water, and the same strategy has been described in a patent relating to processing scrap components containing GaAs where the Ga^{3+} is complexed with a hydroxamic acid, "HA" e.g.:

$$GaAs + H_2O_2 + "HA" \rightarrow "GaA" \text{ complex} + H_3AsO_4 + 4H_2O$$

Oxidation of GaAs occurs in air and degrades performance of the semiconductor, the surface can be passivated by depositing a cubic gallium(II) sulfide layer using a tert-butyl gallium sulfide compound such as $(^tBuGaS)_7$.

GaAs has some electronic properties which are superior to those of silicon.

Precipitation	Precipitation is the formation of a solid in a solution during a chemical reaction. When the reaction occurs, the solid formed is called the precipitate, and the liquid remaining above the solid is called the supernate. Natural methods of Precipitation include settling or sedimentation, where a solid forms over a period of time due to ambient forces like gravity or centrifugation.
Sodium hydroxide	Sodium hydroxide is a caustic metallic base. Sodium hydroxide forms a strong alkaline solution when dissolved in a solvent such as water. However, only the hydroxide ion is basic.
Bauxite	Bauxite is the most important aluminium ore. It consists largely of the minerals gibbsite $Al(OH)_3$, boehmite γ-AlO(OH), and diaspore α-AlO(OH), together with the iron oxides goethite and hematite, the clay mineral kaolinite and small amounts of anatase TiO_2. It was named after the village Les Baux in southern France, where it was first discovered in 1821 by the geologist Pierre Berthier.
Dow process	The Dow process is the electrolytic method of bromine extraction from brine, and was Herbert Henry Dow"s second revolutionary process for generating bromine commercially. Before Dow got into the bromine business, bromine-rich brine was evaporated by heating with wood scraps and then crystallized sodium chloride was removed. An oxidizing agent was added, and bromine was formed in the solution.

Chapter 21. Chemistry of the Main-Group Elements

Chapter 21. Chemistry of the Main-Group Elements

Electrolysis

In chemistry and manufacturing, Electrolysis is a method of using an electric current to drive an otherwise non-spontaneous chemical reaction. Electrolysis is commercially highly important as a stage in the separation of elements from naturally-occurring sources such as ores using an electrolytic cell.

- 1800 - William Nicholson and Johann Ritter decomposed water into hydrogen and oxygen.
- 1807 - Potassium, Sodium, Barium, Calcium and Magnesium were discovered by Sir Humphry Davy using Electrolysis.
- 1886 - Fluorine was discovered by Henri Moissan using Electrolysis.
- 1886 - Hall-Héroult process developed for making aluminium
- 1890 - Castner-Kellner process developed for making sodium hydroxide

Electrolysis is the passage of an electric current through an ionic substance that is either molten or dissolved in a suitable solvent, resulting in chemical reactions at the electrodes and separation of materials.

The main components required to achieve Electrolysis are:

- A liquid containing mobile ions - an electrolyte
- An external source of direct electric current
- Two solid rods or plates known as electrodes

The components perform the following roles in the Electrolysis process:

- The mobile ions are the carriers of electrical current in the liquid (electrolyte.) If the ions are not mobile, as in a solid salt then Electrolysis cannot occur.

Lithium

Lithium is the chemical element with atomic number 3, and is represented by the symbol Li. It is a soft alkali metal with a silver-white color. Under standard conditions it is the lightest metal and the least dense solid element.

Seawater

Seawater is water from a sea or ocean. On average, Seawater in the world"s oceans has a salinity of about 3.5%. This means that every 1 kg of Seawater has approximately 35 grams of dissolved salts (mostly, but not entirely, the ions of sodium chloride: Na^+, Cl^-.)

Reaction mechanism

In chemistry, a Reaction mechanism is the step by step sequence of elementary reactions by which overall chemical change occurs.

Although only the net chemical change is directly observable for most chemical reactions, experiments can often be designed that suggest the possible sequence of steps in a Reaction mechanism.

A mechanism describes in detail exactly what takes place at each stage of a chemical transformation.

Chapter 21. Chemistry of the Main-Group Elements

Chapter 21. Chemistry of the Main-Group Elements

Chemical bond	A Chemical bond is the physical process responsible for the attractive interactions between atoms and molecules, and that which confers stability to diatomic and polyatomic chemical compounds. The explanation of the attractive forces is a complex area that is described by the laws of quantum electrodynamics. In practice, however, chemists usually rely on quantum theory or qualitative descriptions that are less rigorous but more easily explained to describe Chemical bond ing.
Complex	In chemistry, a Complex is a structure consisting of a central atom or molecule connected to surrounding atoms or molecules. Originally, a Complex implied a reversible association of molecules, atoms, or ions through weak chemical bonds. As applied to coordination chemistry, this meaning has evolved.
Electron	The Electron is a subatomic particle that carries a negative electric charge. It has no known substructure and is believed to be a point particle. An Electron has a mass that is approximately 1836 times less than that of the proton.
Electron configuration	In atomic physics and quantum chemistry, Electron configuration is the arrangement of electrons of an atom, a molecule, or other physical structure. It concerns the way electrons can be distributed in the orbitals of the given system (atomic or molecular for instance.) Like other elementary particles, the electron is subject to the laws of quantum mechanics, and exhibits both particle-like and wave-like nature.
Molecular orbital	In chemistry, a Molecular orbital is a mathematical function that describes the wave-like behavior of an electron in a molecule. This function can be used to calculate chemical and physical properties such as the probability of finding an electron in any specific region. The use of the term "orbital" was first used in English by Robert S. Mulliken in 1925 as the English translation of Schrödinger"s use of the German word, "Eigenfunktion".
Molecular orbital theory	In chemistry, Molecular orbital theory is a method for determining molecular structure in which electrons are not assigned to individual bonds between atoms, but are treated as moving under the influence of the nuclei in the whole molecule. In this theory, each molecule has a set of molecular orbitals, in which it is assumed that the molecular orbital wave function ψ_f may be written as a simple weighted sum of the n constituent atomic orbitals χ_i, according to the following equation: $$\psi_j = \sum_{i=1}^{n} c_{ij} \chi_i$$ The c_{ij} coefficients may be determined numerically by substitution of this equation into the Schrödinger equation and application of the variational principle. This method is called the linear combination of atomic orbitals approximation and is used in computational chemistry.

Chapter 21. Chemistry of the Main-Group Elements

Chapter 21. Chemistry of the Main-Group Elements

Zinc	Zinc is a metallic chemical element with the symbol Zn and atomic number 30. It is a first-row transition metal in group 12 of the periodic table. Zinc is chemically similar to magnesium because its ion is of similar size and its only common oxidation state is +2.
Benzene	Benzene, or benzol, is an organic chemical compound and a known carcinogen with the molecular formula C_6H_6. It is sometimes abbreviated Ph-H. Benzene is a colorless and highly flammable liquid with a sweet smell and a relatively high melting point. Because it is a known carcinogen, its use as an additive in gasoline is now limited, but it is an important industrial solvent and precursor in the production of drugs, plastics, synthetic rubber, and dyes.
Configuration	The configuration of a molecule is the permanent geometry that results from the spatial arrangement of its bonds. The ability of the same set of atoms to form two or more molecules with different configurations is stereoisomerism. configuration is distinct from chemical conformation, a shape attainable by bond rotations.
Intermolecular forces	In physics, chemistry, and biology, Intermolecular forces are forces that act between stable molecules or between functional groups of macromolecules. Intermolecular forces include momentary attractions between molecules, diatomic free elements, and individual atoms. They differ from covalent and ionic bonding in that they are not stable, but are caused by momentary polarization of particles.
Ion	An Ion is an atom or molecule where the total number of electrons is not equal to the total number of protons, giving it a net positive or negative electrical charge. Since protons are positively charged and electrons are negatively charged, if there are more electrons than protons, the atom or molecule will be negatively charged. This is called an an Ion , from the Greek á¼€vÎ¬ , meaning "up".
Solid oxide electrolyser cell	A Solid oxide electrolyser cell is a solid oxide fuel cell set in regenerative mode for the electrolysis of water with a solid oxide electrolyte to produce oxygen and hydrogen gas. Solid oxide electrolyser cell s operate at temperatures for high-temperature electrolysis, typically between 500 and 850°C similar to SOFC. Advantages of this class of regenerative fuel cells include high efficiencies, long term stability, fuel flexibility, low emissions, and cost. The largest disadvantage is the high operating temperature which results in longer start up times and mechanical/chemical compatibility issues.
Helium	Helium is the chemical element with atomic number 2, and is represented by the symbol He. It is a colorless, odorless, tasteless, non-toxic, inert monatomic gas that heads the noble gas group in the periodic table. Its boiling and melting points are the lowest among the elements and it exists only as a gas except in extreme conditions.
Magnesium diboride	Magnesium diboride is an inexpensive and simple superconductor. Its superconductivity was announced in the journal Nature in March 2001. Its critical temperature (T_c) of 39 K (−234 °C; −389 °F) is the highest amongst conventional superconductors.

Chapter 21. Chemistry of the Main-Group Elements

Chapter 21. Chemistry of the Main-Group Elements

Niobium	Niobium is the chemical element with the symbol Nb and the atomic number 41. A rare, soft, grey, ductile transition metal, Niobium is found in the minerals pyrochlore, the main commercial source for Niobium, and columbite. Niobium has physical and chemical properties similar to those of the element tantalum, and the two are therefore difficult to distinguish.
Crystal	A Crystal or Crystal line solid is a solid material whose constituent atoms, molecules, or ions are arranged in an orderly repeating pattern extending in all three spatial dimensions. The scientific study of Crystal s and Crystal formation is Crystal lography. The process of Crystal formation via mechanisms of Crystal growth is called Crystal lization or solidification.
Crystal structure	In mineralogy and crystallography, a Crystal structure is a unique arrangement of atoms in a crystal. A Crystal structure is composed of a motif, a set of atoms arranged in a particular way, and a lattice. Motifs are located upon the points of a lattice, which is an array of points repeating periodically in three dimensions.
Liquid Helium	Helium exists in liquid form only at extremely low temperatures The density of Liquid helium at its boiling point and 1 atm is approximately 0.125 g/mL Liquid helium in a cup. Helium-4 was first liquefied on 10 July 1908 by Dutch physicist Heike Kamerlingh Onnes.
Resonance	Resonance in chemistry is a key component of valence bond theory used to graphically represent and mathematically model certain types of molecular structures when no single, conventional Lewis structure can satisfactorily represent the observed structure or explain its properties. Resonance instead considers such molecules to be an intermediate or average (called a Resonance hybrid) between several Lewis structures that differ only in the placement of the valence electrons. Scheme 1.
Lithium hydroxide	Lithium hydroxide is a corrosive alkali hydroxide. It is a white hygroscopic crystalline material. It is soluble in water, and slightly soluble in ethanol.
Calcium hydroxide	Calcium hydroxide, traditionally called slaked lime, hydrated lime, slack lime or pickling lime, is a chemical compound with the chemical formula $Ca(OH)_2$. It is a colourless crystal or white powder, and is obtained when calcium oxide (called lime or quicklime) is mixed, or "slaked" with water. It can also be precipitated by mixing an aqueous solution of calcium chloride and an aqueous solution of sodium hydroxide.
Potassium	Potassium is a chemical element. It has the symbol K , atomic number 19, and atomic mass 39.0983. Potassium was first isolated from potash.
Nitrogen	Nitrogen is a chemical element that has the symbol N and atomic number 7 and atomic mass 14.00674 u. Elemental Nitrogen is a colorless, odorless, tasteless and mostly inert diatomic gas at standard conditions, constituting 78% by volume of Earth"s atmosphere.

Chapter 21. Chemistry of the Main-Group Elements

Chapter 21. Chemistry of the Main-Group Elements

Many industrially important compounds, such as ammonia, nitric acid, organic nitrates, and cyanides, contain Nitrogen.

Reactivity

Reactivity refers to the rate at which a chemical substance tends to undergo a chemical reaction in time. In pure compounds, Reactivity is regulated by the physical properties of the sample. For instance, grinding a sample to a higher specific surface area increases its Reactivity.

Titanium

Titanium is a chemical element with the symbol Ti and atomic number 22. Sometimes called the "space age metal", it has a low density and is a strong, lustrous, corrosion-resistant transition metal with a silver color. Titanium can be alloyed with iron, aluminium, vanadium, molybdenum, among other elements, to produce strong lightweight alloys for aerospace (jet engines, missiles, and spacecraft), military, industrial process (chemicals and petro-chemicals, desalination plants, pulp, and paper), automotive, agri-food, medical prostheses, orthopedic implants, dental and endodontic instruments and files, dental implants, sporting goods, jewelry, mobile phones, and other applications.

Covalent radius

The Covalent radius, r_{cov}, is a measure of the size of atom that forms part of a covalent bond. It is measured either in picometres (pm) or ångströms (Å), with 1 Å = 100 pm.
In principle, the sum of the two covalent radii should equal the covalent bond length between two atoms.

Lye

Lye is a corrosive alkaline substance, commonly, sodium hydroxide Previously, Lye was among the many different alkalis leached from hardwood ashes. Now, Lye is commercially manufactured using a membrane cell method, which is an improvement from the previous diaphragm cell methods of Castner-Kellner, Gibbs, and Nelson.

Soap

Soap is an anionic surfactant used in conjunction with water for washing and cleaning, which historically comes either in solid bars or in the form of a viscous liquid.
Soap consists of sodium or potassium salts of fatty acids and is obtained by reacting common oils or fats with a strong alkaline solution (the base, popularly referred to as lye) in a process known as saponification. The fats are hydrolyzed by the base, yielding alkali salts of fatty acids (crude Soap) and glycerol.

Solvay process

The Solvay process, also referred to as the ammonia-soda process, is the major industrial process for the production of soda ash (sodium carbonate.) The ammonia-soda process was developed into its modern form by Ernest Solvay during the 1860s. The ingredients for this process are readily available and inexpensive: salt brine (from inland sources or from the sea) and limestone (from mines.)

Chapter 21. Chemistry of the Main-Group Elements

Chapter 21. Chemistry of the Main-Group Elements

Colloid	A Colloid is a type of chemical mixture where one substance is dispersed evenly throughout another. The particles of the dispersed substance are only suspended in the mixture, unlike a solution, where they are completely dissolved within. This occurs because the particles in a Colloid are larger than in a solution - small enough to be dispersed evenly and maintain a homogenous appearance, but large enough to scatter light and not dissolve.
Solution	In chemistry, a Solution is a homogeneous mixture composed of two or more substances. In such a mixture, a solute is dissolved in another substance, known as a solvent. Gases may dissolve in liquids, for example, carbon dioxide or oxygen in water.
Beryllium	Beryllium is the chemical element with the symbol Be and atomic number 4. A bivalent element, Beryllium is found naturally only combined with other elements in minerals. Notable gemstones which contain Beryllium include Beryl (aquamarines and emeralds) and Chrysoberyl (Alexandrite and Cat"s eye.)
Potassium hydroxide	Potassium hydroxide is the inorganic compound with the formula KOH. Along with sodium hydroxide, this colourless solid is a prototypical "strong base". It has many industrial and niche applications. Most applications exploit its reactivity toward acids and its corrosive nature.
Potassium nitrate	Potassium nitrate is a chemical compound with the chemical formula KNO_3. A naturally occurring mineral source of nitrogen, KNO_3 constitutes a critical oxidizing component of black powder/gunpowder. In the past it was also used for several kinds of burning fuses, including slow matches.
DNA	Deoxyribonucleic acid (DNA) is a nucleic acid that contains the genetic instructions used in the development and functioning of all known living organisms and some viruses. The main role of DNA molecules is the long-term storage of information. DNA is often compared to a set of blueprints or a recipe, or a code, since it contains the instructions needed to construct other components of cells, such as proteins and RNA molecules.
Magnesium hydroxide	Magnesium hydroxide is an inorganic compound with the chemical formula $Mg(OH)_2$. As a suspension in water, it may be referred to as Milk of Magnesia. The solid mineral form of Magnesium hydroxide is known as brucite.
Milk of magnesia	Milk of Magnesia is an aqueous suspension of magnesium hydroxide, $Mg(OH)_2$ in water. Milk of Magnesia is a saline osmotic (hydrating) laxative. The name derives from the suspension"s milky white appearance and the magnesium in its composition.
Acid rain	Acid rain is rain or any other form of precipitation that is unusually acidic. It has harmful effects on plants, aquatic animals, and infrastructure. Acid rain is mostly caused by human emissions of sulfur and nitrogen compounds which react in the atmosphere to produce acids.

Chapter 21. Chemistry of the Main-Group Elements

Chapter 21. Chemistry of the Main-Group Elements

Coral Calcium	Coral calcium is a salt of calcium derived from fossilized coral reefs. Living coral reefs are endangered and cannot be harvested without significant damage to the ecosystem, and because of this, Coral calcium is harvested by grinding up above-ground limestone deposits that were once part of a coral reef. Calcium from coral sources needs to be refined to remove pollutants of the source environment.
Iron	Iron is a chemical element with the symbol Fe and atomic number 26. Iron is a group 8 and period 4 element. Iron and Iron alloys (steels) are by far the most common metals and the most common ferromagnetic materials in everyday use.
Boron	Boron is the chemical element with atomic number 5 and the chemical symbol B. Boron is a trivalent metalloid element which occurs abundantly in the evaporite ores borax and ulexite. Several allotropes of Boron exist; amorphous Boron is a brown powder, though crystalline Boron is black, extremely hard , and a poor conductor at room temperature. Elemental Boron is used as a dopant in the semiconductor industry, while Boron compounds play important roles as light structural materials, insecticides and preservatives, and reagents for chemical synthesis.
Indium	Indium is a chemical element with chemical symbol In and atomic number 49. This rare, soft, malleable and easily fusible post-transition metal is chemically similar to aluminium or gallium but more closely resembles zinc . Indium"s current primary application is to form transparent electrodes from Indium tin oxide in liquid crystal displays, and this use largely determines its global mining production.
Thallium	Thallium is a chemical element with the symbol Tl and atomic number 81. This soft gray malleable poor metal resembles tin but discolors when exposed to air. Approximately 60-70% of Thallium production is used in the electronics industry, and the rest is used in the pharmaceutical industry and in glass manufacturing.
Tin	Tin is a chemical element with the symbol Sn and atomic number 50. It is a main group metal in group 14 of the periodic table. Tin shows chemical similarity to both neighboring group 14 elements, germanium and lead, like the two possible oxidation states +2 and +4.
Catalysis	Catalysis is the process in which the rate of a chemical reaction is either increased or decreased by means of a chemical substance known as a catalyst. Unlike other reagents that participate in the chemical reaction, a catalyst is not consumed by the reaction itself. The catalyst may participate in multiple chemical transformations.
Heterogeneous catalysis	Heterogeneous catalysis is a chemistry term which describes catalysis where the catalyst is in a different phase (ie. solid, liquid and gas, but also oil and water) to the reactants. Heterogeneous catalysts provide a surface on which the reaction may take place.
Thermite	Thermite is a pyrotechnic composition of a metal powder and a metal oxide, which produces an aluminothermic reaction known as a Thermite reaction. It is not explosive, but can create short bursts of extremely high temperatures focused on a very small area for a short period of time.

Chapter 21. Chemistry of the Main-Group Elements

Chapter 21. Chemistry of the Main-Group Elements

	Thermite s can be a diverse class of compositions.
Atomic mass	The Atomic mass is the mass of an atom, most often expressed in unified Atomic mass units. The Atomic mass may be considered to be the total mass of protons, neutrons and electrons in a single atom (when the atom is motionless.) The Atomic mass is sometimes incorrectly used as a synonym of relative Atomic mass, average Atomic mass and atomic weight; however, these differ subtly from the Atomic mass.
Freezing	In physical science, Freezing or solidification is the process in which a liquid turns into a solid when cold enough. The Freezing point is the temperature at which this happens. Melting, the process of turning a solid to a liquid, is almost the exact opposite of Freezing.
Amphoteric	In chemistry, an amphoteric substance is one that can react as either an acid or base. Many metals (such as zinc, tin, lead, aluminium, and beryllium) and most metalloids have amphoteric oxides or hydroxides. Another class of amphoteric substances are amphiprotic molecules which can either donate or accept a proton.
Lead	Lead is a main-group element with symbol Pb and atomic number 82. Lead is a soft, malleable poor metal, also considered to be one of the heavy metals. Lead has a bluish-white color when freshly cut, but tarnishes to a dull grayish color when exposed to air.
Bromine	Bromine , Greek: βρά¿¶μος, brómos, meaning "stench "), is a chemical element with the symbol Br and atomic number 35. A halogen element, Bromine is a reddish-brown volatile liquid at standard room temperature that is intermediate in reactivity between chlorine and iodine. Bromine vapours are corrosive and toxic.
White phosphorus	White phosphorus is a flare- and smoke-producing agent and an incendiary agent that is made from a common allotrope of the chemical element phosphorus. The main utility of White phosphorus munitions is to create smokescreens to mask movement from the enemy, or to mask his fire. In contrast to other smoke-causing munitions, White phosphorus burns quickly causing an instant bank of smoke.
Phosphorus	Phosphorus is the chemical element that has the symbol P and atomic number 15. A multivalent nonmetal of the nitrogen group, Phosphorus is commonly found in inorganic phosphate rocks. Elemental Phosphorus exists in two major forms - white Phosphorus and red Phosphorus.

Chapter 21. Chemistry of the Main-Group Elements

Chapter 21. Chemistry of the Main-Group Elements

Deuterium	Deuterium, also called heavy hydrogen, is a stable isotope of hydrogen with a natural abundance in the oceans of Earth of approximately one atom in 6,500 of hydrogen (~154 ppm.) Deuterium thus accounts for approximately 0.015% (alternately, on a weight basis: 0.031%) of all naturally occurring hydrogen in the oceans on Earth Deuterium abundance on Jupiter is about 2.25×10^{-5} (roughly 22 atoms in a million, or 15% of the terrestrial Deuterium-to-hydrogen ratio); these ratios presumably reflect the early solar nebula ratios, and those after the Big Bang.
Tritium	Tritium is a radioactive isotope of hydrogen. The nucleus of Tritium contains one proton and two neutrons, whereas the nucleus of protium (the most abundant hydrogen isotope) contains one proton and no neutrons. While Tritium has several different experimentally-determined values of its half-life, the NIST recommends 4,500±8 days (approximately 12.33 years.)
Ionization	Ionization is the physical process of converting an atom or molecule into an ion by adding or removing charged particles such as electrons or other ions. This is often confused with dissociation (chemistry.) The process works slightly differently depending on whether an ion with a positive or a negative electric charge is being produced.
Isotopes	Isotopes are any of the different types of atoms of the same chemical element, each having a different atomic mass (mass number.) Isotopes of an element have nuclei with the same number of protons (the same atomic number) but different numbers of neutrons. Therefore, Isotopes of the same element have different mass numbers (number of nucleons.)
Ammonia	Ammonia is a compound of nitrogen and hydrogen with the formula NH_3. It is normally encountered as a gas with a characteristic pungent odor. Ammonia contributes significantly to the nutritional needs of terrestrial organisms by serving as a precursor to foodstuffs and fertilizers.
Cobalt	Cobalt is a hard, lustrous, grey metal, a chemical element with symbol Co and atomic number 27. Although Cobalt based colors and pigments have been used since ancient times for making jewelry and paints, and miners have long used the name kobold ore for some minerals, the free metallic Cobalt was not prepared and discovered until 1735 by Georg Brandt.
Hydrogenation	Hydrogenation is the chemical reaction that results from the addition of hydrogen (H_2.) The process is usually employed to reduce or saturate organic compounds. The process typically constitutes the addition of pairs of hydrogen atoms to a molecule.
Tungsten	Tungsten is a chemical element with the chemical symbol W and atomic number 74.

Chapter 21. Chemistry of the Main-Group Elements

Chapter 21. Chemistry of the Main-Group Elements

	A steel-gray metal, Tungsten is found in several ores, including wolframite and scheelite. It is remarkable for its robust physical properties, especially the fact that it has the highest melting point of all the non-alloyed metals and the second highest of all the elements after carbon.
Aqueous solution	An Aqueous solution is a solution in which the solvent is water. It is usually shown in chemical equations by appending (aq) to the relevant formula. The word aqueous means pertaining to, related to, similar to, or dissolved in water.
Catalyst	Catalyst is a science centre and museum devoted to the chemical industry. Its full title is Catalyst Science Discovery Centre. It is located in Widnes, Cheshire, in the north west of England, and situated on the north bank of the River Mersey (grid reference SJ512841.)
Reducing agent	A Reducing agent is the element or compound in a redox reaction that reduces another species. In doing so, it becomes oxidized, and is therefore the electron donor in the redox. For example consider the following reaction: $[Fe_6]^{4-} + 1/2\ Cl_2 \rightarrow [Fe_6]^{3-} + Cl^-$ The Reducing agent in this reaction is ferrocyanide: it donates an electron, converting to ferricyanide, simultaneous with the reduction of chlorine to chloride.
Carbon black	Carbon black is a material produced by the incomplete combustion of heavy petroleum products such as FCC tar, coal tar, ethylene cracking tar, and a small amount from vegetable oil. Carbon black is a form of amorphous carbon that has a high surface area to volume ratio, although its surface area to volume ratio is low compared to activated carbon. It is dissimilar to soot because of its much higher surface area to volume ratio and significantly less (negligible and non-bioavailable) PAH content.
Catenation	Catenation is the ability of a chemical element to form a long chain-like structure via a series of covalent bonds. Catenation occurs most readily in carbon, which forms covalent bonds with other carbon atoms.Catenation is the reason for the presence of large number of organic compounds in nature. Carbon is most well known for its properties of Catenation, with organic chemistry essentially being the study of catenated carbon structures (otherwise known as catenae.)
Diamond	In mineralogy, Diamond is an allotrope of carbon, where the carbon atoms are arranged in a variation of the face centered cubic crystal structure called a Diamond lattice. Diamond is the second most stable form of carbon, after graphite; however, the conversion rate from Diamond to graphite is negligible at ambient conditions. Diamond is specifically renowned as a material with superlative physical qualities, most of which originate from the strong covalent bonding between its atoms.

Chapter 21. Chemistry of the Main-Group Elements

Chapter 21. Chemistry of the Main-Group Elements

Hydrocarbon	In organic chemistry, a Hydrocarbon is an organic compound consisting entirely of hydrogen and carbon. With relation to chemical terminology, aromatic Hydrocarbon s or arenes, alkanes, alkenes and alkyne-based compounds composed entirely of carbon or hydrogen are referred to as "pure" Hydrocarbon s, whereas other Hydrocarbon s with bonded compounds or impurities of sulfur or nitrogen, are referred to as "impure", and remain somewhat erroneously referred to as Hydrocarbon s. Hydrocarbon s are referred to as consisting of a "backbone" or "skeleton" composed entirely of carbon and hydrogen and other bonded compounds, and have a functional group that generally facilitates combustion.
Carbonic acid	Carbonic acid (ancient name acid of air or aerial acid) has the formula H_2CO_3. It is also a name sometimes given to solutions of carbon dioxide in water, which contain small amounts of H_2CO_3. The salts of Carbonic acid s are called bicarbonates (or hydrogen carbonates) and carbonates.
Mole	The mole is a unit of amount of substance: it is an SI base unit, and one of the few units used to measure this physical quantity. The name "mole" was coined in German by Wilhelm Ostwald in 1893, although the related concept of equivalent mass had been in use at least a century earlier. The name is assumed to be derived from the word Molekül (molecule.)
Condensation	Condensation is the change of the physical state of aggregation (or simply state) of matter from gaseous phase into liquid phase. When the transition happens from the gaseous phase into the solid phase directly, bypassing the liquid phase, the change is called deposition. Condensation commonly occurs when a vapor is cooled to its dew point, but the dew point can also be reached through compression.
Condensation reaction	A Condensation reaction is a chemical reaction in which two molecules or moieties (functional groups) combine to form one single molecule, together with the loss of a small molecule. When this small molecule is water, it is known as a dehydration reaction; other possible small molecules lost are hydrogen chloride, methanol, or acetic acid. The word "condensation" suggests a process in which something is lost; for reactions a small molecule is lost.
Silicic acid	Silicic acid is a general name for a family of chemical compounds of the element silicon, hydrogen, and oxygen, with the general formula $[SiO_x(OH)_{4-2x}]_n$. Some simple Silicic acid s have been identified in very dilute aqueous solution, such as meta Silicic acid (H_2SiO_3), ortho Silicic acid (H_4SiO_4, pK_{a1}=9.84, pK_{a2}=13.2 at 25°C), di Silicic acid ($H_2Si_2O_5$), and pyro Silicic acid ($H_6Si_2O_7$); however in the solid state these probably condense to form polymeric Silicic acid s of complex structure. Silicic acid s may be formed by acidification of silicate salts (such as sodium silicate) in aqueous solution.
Antimony	Antimony) is a chemical element with the symbol Sb and atomic number 51. A metalloid, Antimony has four allotropic forms. The stable form of Antimony is a blue-white metalloid.

Chapter 21. Chemistry of the Main-Group Elements

Chapter 21. Chemistry of the Main-Group Elements

Arsenic	Arsenic is the chemical element that has the symbol As and atomic number 33. Arsenic was first documented by Albertus Magnus in 1250. Its atomic mass is 74.92.
Bismuth	Bismuth is a chemical element that has the symbol Bi and atomic number 83. This trivalent poor metal chemically resembles arsenic and antimony. Bismuth is heavy and brittle; it has a silvery white color with a pink tinge due to the surface oxide.
Polymer	A Polymer is a large molecule composed of repeating structural units typically connected by covalent chemical bonds. While Polymer in popular usage suggests plastic, the term actually refers to a large class of natural and synthetic materials with a variety of properties. Due to the extraordinary range of properties accessible in Polymer ic materials , they have come to play an essential and ubiquitous role in everyday life - from plastics and elastomers on the one hand to natural bio Polymer s such as DNA and proteins that are essential for life on the other.
Electron affinity	The Electron affinity, E_{ea}, of an atom or molecule is the amount of energy required to detach an electron from a singly charged negative ion, i.e., the energy change for the process $$X^- \rightarrow X + e^-$$ An equivalent definition is the energy released ($E_{initial} - E_{final}$) when an electron is attached to a neutral atom or molecule. All elements whose Electron affinity have been measured using modern methods have a positive Electron affinity, but older texts mistakenly report that some elements such as alkaline earth metals have negative E_{ea}, meaning they would repel electrons. This is not recognized by modern chemists.
Nitrogen Group	The Nitrogen group is Group 15 of the periodic table (formerly numbered as Group V or Group VA) and is also collectively named the pnictogens. This consists of nitrogen (N), phosphorus (P), arsenic (As), antimony (Sb), bismuth (Bi) and ununpentium (Uup) (unconfirmed.) Like other groups, the members of this family show patterns in its electron configuration, especially the outermost shells resulting in trends in chemical behavior: This group has the defining characteristic that all the component elements have 5 electrons in their outermost shell, that is 2 electrons in the s subshell and 3 unpaired electrons in the p subshell.
Nitrogen cycle	The Nitrogen cycle is the biogeochemical cycle that describes the transformations of nitrogen and nitrogen-containing compounds in nature. It is a cycle which includes gaseous components. Earth"s atmosphere is approximately 78-80% nitrogen, making it the largest pool of nitrogen.
Rutherford	The Rutherford is an obsolete unit of radioactivity, defined as the activity of a quantity of radioactive material in which one million nuclei decay per second. It is therefore equivalent to one megabecquerel. It was named after Ernest Rutherford It is not an SI unit.

Chapter 21. Chemistry of the Main-Group Elements

Chapter 21. Chemistry of the Main-Group Elements

Haber process	The Haber process is the nitrogen fixation reaction of nitrogen gas and hydrogen gas, over an enriched iron catalyst, to produce ammonia. The Haber process is important because ammonia is difficult to produce on an industrial scale, and the fertilizer generated from the ammonia is responsible for sustaining one-third of the Earth"s population. Despite the fact that 78.1% of the air we breathe is nitrogen, the gas is relatively unreactive because nitrogen molecules are held together by strong triple bonds.
Liquid nitrogen	Liquid nitrogen is a liquefied atmospheric gas produced industrially in large quantities by fractional distillation of liquid air. It is pure nitrogen in a liquid state at a very low temperature. Liquid nitrogen is a colourless clear liquid with density at its boiling point of 0.807 g/mL and a dielectric constant of 1.4.
Uranium	Uranium is a silvery-white metallic chemical element in the actinide series of the periodic table that has the symbol U and atomic number 92. Besides its 92 protons, a Uranium nucleus can have between 141 and 146 neutrons, with 146 and 143 (U-235) in its most common isotopes. The number of electrons in a Uranium atom is 92, 6 of them valence electrons.
Molecular model	A Molecular model in this article, is a physical model that represents molecules and their processes. The creation of mathematical models of molecular properties and behaviour is Molecular model ling, and their graphical depiction is molecular graphics, but these topics are closely linked and each uses techniques from the others. In this article, Molecular model will primarily refer to systems containing more than one atom and where nuclear structure is neglected.
Nitric acid	Nitric acid, also known as aqua fortis and spirit of nitre, is a highly corrosive and toxic strong acid that can cause severe burns. Colorless when pure, older samples tend to acquire a yellow cast due to the accumulation of oxides of nitrogen. If the solution contains more than 86% Nitric acid, it is referred to as fuming Nitric acid.
Ostwald process	The Ostwald process is a chemical process for producing nitric acid, which was developed by Wilhelm Ostwald (patented 1902.) It is a mainstay of the modern chemical industry. Historically and practically it is closely associated with the Haber process, which provides the requisite raw material, ammonia.
Phosphorous acid	Phosphorous acid is the compound described by the formula H_3PO_3. It can be formulated as HP(O)(OH)$_2$ and therefore contains phosphorus in oxidation state +3. It is one of the oxoacids of phosphorus, other important members being phosphoric acid (H_3PO_4) and hypo Phosphorous acid (H_3PO_2.)

Chapter 21. Chemistry of the Main-Group Elements

Chapter 21. Chemistry of the Main-Group Elements

Oxoacid	An Oxoacid is an acid which contains oxygen. More specifically, it is an acid which: 1. contains oxygen; 2. contains at least one other element; 3. has at least one hydrogen atom bound to oxygen; and 4. forms an ion by the loss of one or more protons. The name oxyacid is sometimes used, although this is not recommended. Generally, Oxoacid s are simply polyatomic ions with a hydrogen cation. Under Lavoisier"s original theory, all acids contained oxygen, which was named from the Greek οξυς (acid, sharp) and γεινομαι (geinomai) (engender.)
Halide	A Halide is a binary compound, of which one part is a halogen atom and the other part is an element or radical that is less electronegative than the halogen, to make a fluoride, chloride, bromide, iodide, or astatide compound. Many salts are Halide s. All Group 1 metals form Halide s with the halogens and they are white solids.
Halogen	The Halogen s or Halogen elements are a series of nonmetal elements from Group 17 IUPAC Style (formerly: VII, VIIA) of the periodic table, comprising fluorine, (F); chlorine, (Cl); bromine, (Br); iodine, (I); and astatine, (At.) The undiscovered element 117, temporarily named ununseptium, may also be a Halogen The group of Halogen s is the only periodic table group which contains elements in all three familiar states of matter at standard temperature and pressure.
Detergent	A Detergent is a material intended to assist cleaning. The term is sometimes used to differentiate between soap and other surfactants used for cleaning. As an adjective pertaining to a substance, it (or "detersive") means "cleaning" or "having cleaning properties"; "detergency" indicates presence or degree of cleaning property.
Polonium	Polonium is a chemical element with the symbol Po and atomic number 84, discovered in 1898 by Marie and Pierre Curie. A rare and highly radioactive metalloid, Polonium is chemically similar to bismuth and tellurium, and it occurs in uranium ores. Polonium has been studied for possible use in heating spacecraft.
Selenium	Selenium is a chemical element with the atomic number 34, represented by the chemical symbol Se, an atomic mass of 78.96. It is a nonmetal, chemically related to sulfur and tellurium, and rarely occurs in its elemental state in nature. Isolated Selenium occurs in several different forms, the most stable of which is a dense purplish-gray semi-metal form that is structurally a trigonal polymer chain.
Tellurium	Tellurium is a chemical element that has the symbol Te and atomic number 52. A brittle silver-white metalloid which looks like tin, Tellurium is chemically related to selenium and sulfur. Tellurium is primarily used in alloys and as a semiconductor.

Chapter 21. Chemistry of the Main-Group Elements

Chapter 21. Chemistry of the Main-Group Elements

Eutrophication	Eutrophication is an increase in chemical nutrients -- compounds containing nitrogen or phosphorus -- in an ecosystem, and may occur on land or in water. However, the term is often used to mean the resultant increase in the ecosystem"s primary productivity (excessive plant growth and decay), and further effects including lack of oxygen and severe reductions in water quality, fish, and other animal populations. The Eutrophication of the Potomac River is evident from its bright green water, caused by a dense bloom of cyanobacteria.
	Eutrophication is frequently a result of nutrient pollution, such as the release of sewage effluent, urban stormwater run-off, and run-off carrying excess fertilizers into natural waters.
Dioxygen	Dioxygen plays an important role in the energy metabolism of living organisms. Free oxygen is produced in the biosphere through photolysis (light-driven oxidation and splitting) of water during photosynthesis in cyanobacteria, green algae, and plants. During oxidative phosphorylation in cellular respiration, oxygen is reduced to water, thus closing the biological water-oxygen redox cycle.
Manganese	Manganese is a chemical element, designated by the symbol Mn. It has the atomic number 25. It is found as a free element in nature , and in many minerals.
Joseph Priestley	Joseph Priestley - 6 February 1804) was an 18th-century British theologian, Dissenting clergyman, natural philosopher, educator, and political theorist who published over 150 works. He is usually credited with the discovery of oxygen, having isolated it in its gaseous state, although Carl Wilhelm Scheele and Antoine Lavoisier also have a claim to the discovery.
	During his lifetime, Priestley"s considerable scientific reputation rested on his invention of soda water, his writings on electricity, and his discovery of several "airs" , the most famous being what Priestley dubbed "dephlogisticated air" (oxygen.)
Monoclinic	In crystallography, the monoclinic crystal system is one of the 7 lattice point groups. A crystal system is described by three vectors. In the monoclinic system, the crystal is described by vectors of unequal length, as in the orthorhombic system.
Claus process	The Claus process is the most significant gas desulfurizing process, recovering elemental sulfur from gaseous hydrogen sulfide. First patented in 1883 by the scientist Carl Friedrich Claus, the Claus process has become the industry standard.
	The multi-step Claus process recovers sulfur from the gaseous hydrogen sulfide found in raw natural gas and from the by-product gases containing hydrogen sulfide derived from refining crude oil and other industrial processes.
Hydrogen sulfide	Hydrogen sulfide is the chemical compound with the formula H_2S. This colorless, toxic and flammable gas is partially responsible for the foul odor of rotten eggs and flatulence.
	It often results from the bacterial break down of sulfites in nonorganic matter in the absence of oxygen, such as in swamps and sewers (anaerobic digestion.) It also occurs in volcanic gases, natural gas and some well waters.

Chapter 21. Chemistry of the Main-Group Elements

Chapter 21. Chemistry of the Main-Group Elements

Boiling	Boiling, a type of phase transition, is the rapid vaporization of a liquid, which typically occurs when a liquid is heated to its Boiling point, the temperature at which the vapor pressure of the liquid is equal to the pressure exerted on the liquid by the surrounding environmental pressure. Thus, a liquid may also boil when the pressure of the surrounding atmosphere is sufficiently reduced, such as the use of a vacuum pump or at high altitudes. Boiling occurs in three characteristic stages, which are nucleate, transition and film Boiling.
Melting	Melting is a physical process that results in the phase change of a substance from a solid to a liquid. The internal energy of a solid substance is increased, typically by the application of heat or pressure, resulting in a rise of its temperature to the Melting point, at which the rigid ordering of molecular entities in the solid breaks down to a less-ordered state and the solid liquefies. An object that has melted completely is molten.
Sulfur Oxoacids	The Sulfur oxoacids are chemical compounds that contain sulfur, oxygen and hydrogen. The best known and most important industrially is sulfuric acid)
Bleach	A bleach is a chemical that removes colors or whitens, often via oxidation. Common chemical bleach es include household "chlorine bleach , a solution of approximately 3-6% sodium hypochlorite (NaClO), and "oxygen bleach , which contains hydrogen peroxide or a peroxide-releasing compound such as sodium perborate, sodium percarbonate, sodium persulfate, sodium perphosphate, or urea peroxide together with catalysts and activators, e.g. tetraacetylethylenediamine and/or sodium nonanoyloxybenzenesulfonate. To bleach something is to apply bleach sometimes as a preliminary step in the process of dyeing.
Contact process	The Contact process is the current method of producing sulfuric acid in the high concentrations needed for industrial processes. Vanadium(V) oxide (vanadium pentoxide) is the catalyst employed. This process was patented in 1831 by the British vinegar merchant Peregrine Phillips.
Astatine	Astatine is a radioactive chemical element with the symbol At and atomic number 85. It is the heaviest of the discovered halogens. Although Astatine is produced by radioactive decay in nature, due to its short half life it is found only in minute amounts.
Calcium fluoride	Calcium fluoride is an insoluble ionic compound of calcium and fluorine. It occurs naturally as the mineral fluorite (also called fluorspar) and as Blue John, and it is the source of most of the world"s fluorine. This insoluble solid adopts a cubic structure wherein calcium is coordinated to eight fluoride anions and each F^- ion is surrounded by four Ca^{2+} ions.
Fluorine	Fluorine is a chemical element, represented by the symbol F, and the atomic number 9. Fluorine forms a single bond with itself in elemental form, resulting in the diatomic F_2 molecule. F_2 is a supremely reactive, poisonous, pale, yellowish brown gas.

Chapter 21. Chemistry of the Main-Group Elements

Chapter 21. Chemistry of the Main-Group Elements

Hydrochloric acid	Hydrochloric acid is the solution of hydrogen chloride (HCl) in water. It is a highly corrosive, strong mineral acid and has major industrial uses. It is found naturally in gastric acid.
Silver	Silver is a chemical element with the chemical symbol Ag and atomic number 47. A soft, white, lustrous transition metal, it has the highest electrical conductivity of any element and the highest thermal conductivity of any metal. The metal occurs naturally in its pure, free form, as an alloy with gold (electrum) and other metals, and in minerals such as argentite and chlorargyrite.
Chlorin	In organic chemistry, a Chlorin is a large heterocyclic aromatic ring consisting, at the core, of three pyrroles and one pyrroline coupled through four methine linkages. Unlike a porphyrin, a Chlorin is therefore largely aromatic but not aromatic through the entire circumference of the ring. Magnesium-containing Chlorin s are called chlorophylls, and are the central photosensitive pigment in chloroplasts.
Chlorine	Chlorine . As the chloride ion, which is part of common salt and other compounds, it is abundant in nature and necessary to most forms of life, including humans. In its elemental form (Cl_2 or "di Chlorine ") under standard conditions, Chlorine is a powerful oxidant and is used in bleaching and disinfectants.
Sulfuric acid	Sulfuric (or sulphuric) acid, H_2SO_4, is a strong mineral acid. It is soluble in water at all concentrations. Sulfuric acid has many applications, and is one of the top products of the chemical industry.
Hydrogen bromide	Hydrogen bromide is the diatomic molecule Hydrogen bromide r. Hydrogen bromide r is a gas that liquifies upon cooling. Hydrobromic acid forms upon dissolving Hydrogen bromide r in water.
Hydrogen iodide	Hydrogen iodide is a diatomic molecule. Aqueous solutions of Hydrogen iodide are known as iohydroic acid or hydriodic acid, a strong acid. Hydrogen iodide and hydroiodic acid are, however, different in that the former is a gas under standard conditions; whereas, the other is an aqueous solution of said gas.
Chloric acid	Chloric acid, $HClO_3$, is an oxoacid of chlorine, and the formal precursor of chlorate salts. It is a strong acid ($pK_a \approx -1$) and oxidizing agent. It is prepared by the reaction of sulfuric acid with barium chlorate, the insoluble barium sulfate being removed by precipitation: $$Ba(ClO_3)_2 + H_2SO_4 \rightarrow 2HClO_3 + BaSO_4$$

Chapter 21. Chemistry of the Main-Group Elements

Chapter 21. Chemistry of the Main-Group Elements

	Another method is the heating of hypochlorous acid, of which productions include Chloric acid and hydrogen chloride:
	$3HClO \rightarrow HClO_3 + 2\ HCl$
	It is stable in cold aqueous solution up to a concentration of approximately 30%, and solution of up to 40% can be prepared by careful evaporation under reduced pressure.
Perchloric acid	Perchloric acid, $HClO_4$, is an oxoacid of chlorine and is a colorless liquid soluble in water. It is a strong acid comparable in strength to sulfuric and nitric acids. It is useful for preparing perchlorate salts, but it is also dangerously corrosive and readily forms explosive mixtures.
Argon	Argon is a chemical element designated by the symbol Ar. Argon has atomic number 18 and is the third element in group 18 of the periodic table . Argon is present in the Earth"s atmosphere at 0.94%.
Krypton	Krypton is a chemical element with the symbol Kr and atomic number 36. It is a member of Group 18 and Period 4 elements. A colorless, odorless, tasteless noble gas, Krypton occurs in trace amounts in the atmosphere, is isolated by fractionally distilling liquified air, and is often used with other rare gases in fluorescent lamps.
Neon	Neon is the chemical element that has the symbol Ne and atomic number 10. Although a very common element in the universe, it is rare on Earth. A colorless, inert noble gas under standard conditions, Neon gives a distinct reddish-orange glow when used in discharge tubes and Neon lamps.
Noble gas	The Noble gas es are a group of chemical elements with very similar properties: under standard conditions, they are all odorless, colorless, monatomic gases, with a very low chemical reactivity. The six Noble gas es that occur naturally are helium (He), neon (Ne), argon (Ar), krypton (Kr), xenon (Xe), and the radioactive radon (Rn.)
	For the first six periods of the periodic table, the Noble gas es are exactly the members of group 18 of the periodic table.
Platinum	Platinum is a chemical element with the chemical symbol Pt and an atomic number of 78." It is in Group 10 of the periodic table of elements. A dense, malleable, ductile, precious, gray-white transition metal, Platinum is resistant to corrosion and occurs in some nickel and copper ores along with some native deposits. Platinum is used in jewelry, laboratory equipment, electrical contacts and electrodes, Platinum resistance thermometers, dentistry equipment, and catalytic converters.

Chapter 21. Chemistry of the Main-Group Elements

Chapter 21. Chemistry of the Main-Group Elements

Radon	Radon is a chemical element with symbol Rn and atomic number 86. Radon is a colorless, odorless, tasteless, naturally occurring, radioactive noble gas that is formed from the decay of radium. It is one of the heaviest substances that remains a gas under normal conditions and is considered to be a health hazard.
William Ramsay	Sir William Ramsay, KCB (2 October 1852 - 23 July 1916) was a Scottish chemist who discovered the noble gases and received the Nobel Prize in Chemistry in 1904 "in recognition of his services in the discovery of the inert gaseous elements in air" (along with Lord Rayleigh who received the Nobel Prize in Physics that same year for the discovery of argon.)
	Ramsay was born in Glasgow, the son of William Ramsay, C.E. and Catherine, née Robertson. He was a nephew of the geologist Sir Andrew Ramsay.

Chapter 21. Chemistry of the Main-Group Elements

Chapter 22. The Transition Elements and Coordination Compounds

Boiling	Boiling, a type of phase transition, is the rapid vaporization of a liquid, which typically occurs when a liquid is heated to its Boiling point, the temperature at which the vapor pressure of the liquid is equal to the pressure exerted on the liquid by the surrounding environmental pressure. Thus, a liquid may also boil when the pressure of the surrounding atmosphere is sufficiently reduced, such as the use of a vacuum pump or at high altitudes. Boiling occurs in three characteristic stages, which are nucleate, transition and film Boiling.
Cadmium	Cadmium is a chemical element with the symbol Cd and atomic number 48. The soft, bluish-white transition metal is chemically similar to the two other metals in group 12 , zinc and mercury. Similar to zinc it prefers oxidation state +2 in most of its compounds and similar to mercury it shows a low melting point for a transition metal.
Copper	Copper is a chemical element with the symbol Cu and atomic number 29. It is a ductile metal with very high thermal and electrical conductivity. Pure Copper is rather soft and malleable and a freshly-exposed surface has a pinkish or peachy color.
Iron	Iron is a chemical element with the symbol Fe and atomic number 26. Iron is a group 8 and period 4 element. Iron and Iron alloys (steels) are by far the most common metals and the most common ferromagnetic materials in everyday use.
Melting	Melting is a physical process that results in the phase change of a substance from a solid to a liquid. The internal energy of a solid substance is increased, typically by the application of heat or pressure, resulting in a rise of its temperature to the Melting point, at which the rigid ordering of molecular entities in the solid breaks down to a less-ordered state and the solid liquefies. An object that has melted completely is molten.
Melting point	The Melting point of a solid is the temperature range at which it changes state from solid to liquid. At the Melting point the solid and liquid phase exist in equilibrium. When considered as the temperature of the reverse change from liquid to solid, it is referred to as the freezing point.
Periodic trends	In Chemistry, Periodic trends are the tendencies of certain elemental characteristics to increase or decrease as one progresses from one corner of the Periodic table of elements. The atomic radius is the distance from the atomic nucleus to the outermost stable electron orbital in an atom that is at equilibrium. The atomic radius tends to decrease as one progresses across a period because the effective nuclear charge increases, thereby attracting the orbiting electrons and lessening the radius.
Zinc	Zinc is a metallic chemical element with the symbol Zn and atomic number 30. It is a first-row transition metal in group 12 of the periodic table. Zinc is chemically similar to magnesium because its ion is of similar size and its only common oxidation state is +2.

Chapter 22. The Transition Elements and Coordination Compounds

Chapter 22. The Transition Elements and Coordination Compounds

Buffer solution	A Buffer solution is an aqueous solution consisting of a mixture of a weak acid and its conjugate base or a weak base and its conjugate acid. It has the property that the pH of the solution changes very little when a small amount of acid or base is added to it. Buffer solution s are used as a means of keeping pH at a nearly constant value in a wide variety of chemical applications.
Configuration	The configuration of a molecule is the permanent geometry that results from the spatial arrangement of its bonds. The ability of the same set of atoms to form two or more molecules with different configurations is stereoisomerism. configuration is distinct from chemical conformation, a shape attainable by bond rotations.
Earth	Earth is the third planet from the Sun, and the largest of the terrestrial planets in the Solar System in terms of diameter, mass and density. It is also referred to as the World, the Blue Planet, and Terra. Home to millions of species, including humans, Earth is the only place in the universe where life is known to exist.
Electron	The Electron is a subatomic particle that carries a negative electric charge. It has no known substructure and is believed to be a point particle. An Electron has a mass that is approximately 1836 times less than that of the proton.
Electron configuration	In atomic physics and quantum chemistry, Electron configuration is the arrangement of electrons of an atom, a molecule, or other physical structure. It concerns the way electrons can be distributed in the orbitals of the given system (atomic or molecular for instance.) Like other elementary particles, the electron is subject to the laws of quantum mechanics, and exhibits both particle-like and wave-like nature.
Isomer	In chemistry, Isomer s are compounds with the same molecular formula but different structural formula. Isomer s do not necessarily share similar properties unless they also have the same functional groups. This should not be confused with a nuclear Isomer which involves a nucleus at different states of excitement.
Polymer	A Polymer is a large molecule composed of repeating structural units typically connected by covalent chemical bonds. While Polymer in popular usage suggests plastic, the term actually refers to a large class of natural and synthetic materials with a variety of properties. Due to the extraordinary range of properties accessible in Polymer ic materials , they have come to play an essential and ubiquitous role in everyday life - from plastics and elastomers on the one hand to natural bio Polymer s such as DNA and proteins that are essential for life on the other.
Solution	In chemistry, a Solution is a homogeneous mixture composed of two or more substances. In such a mixture, a solute is dissolved in another substance, known as a solvent. Gases may dissolve in liquids, for example, carbon dioxide or oxygen in water.
Beryllium	Beryllium is the chemical element with the symbol Be and atomic number 4.

Chapter 22. The Transition Elements and Coordination Compounds

Chapter 22. The Transition Elements and Coordination Compounds

	A bivalent element, Beryllium is found naturally only combined with other elements in minerals. Notable gemstones which contain Beryllium include Beryl (aquamarines and emeralds) and Chrysoberyl (Alexandrite and Cat"s eye.)
Chromium	Chromium is a chemical element which has the symbol Cr and atomic number 24. It is a steely-gray, lustrous, hard metal that takes a high polish and has a high melting point. It is also odourless, tasteless, and malleable.
Oxidation state	In chemistry, the Oxidation state is an indicator of the degree of oxidation of an atom in a chemical compound. The formal Oxidation state is the hypothetical charge that an atom would have if all bonds to atoms of different elements were 100% ionic. Oxidation state s are typically represented by integers, which can be positive, negative, or zero.
Pauli exclusion principle	The Pauli exclusion principle is a quantum mechanical principle formulated by Wolfgang Pauli in 1925. It states that no two identical fermions may occupy the same quantum state simultaneously. A more rigorous statement of this principle is that, for two identical fermions, the total wave function is anti-symmetric.
Scandium	Scandium is a chemical element with symbol Sc and atomic number 21. A silvery-white metallic transition metal, it has historically been sometimes classified as a rare earth element, together with yttrium and the lanthanides. In 1879 Lars Fredrik Nilson and his team, found a new element with spectral analysis, in the minerals euxenite and gadolinite from Scandinavia.
Titanium	Titanium is a chemical element with the symbol Ti and atomic number 22. Sometimes called the "space age metal", it has a low density and is a strong, lustrous, corrosion-resistant transition metal with a silver color. Titanium can be alloyed with iron, aluminium, vanadium, molybdenum, among other elements, to produce strong lightweight alloys for aerospace (jet engines, missiles, and spacecraft), military, industrial process (chemicals and petro-chemicals, desalination plants, pulp, and paper), automotive, agri-food, medical prostheses, orthopedic implants, dental and endodontic instruments and files, dental implants, sporting goods, jewelry, mobile phones, and other applications.
Vanadium	Vanadium is the chemical element with the symbol V and atomic number 23. It is a soft, silvery grey, ductile transition metal. The formation of an oxide layer stabilizes the metal against oxidation.
Atomic mass	The Atomic mass is the mass of an atom, most often expressed in unified Atomic mass units. The Atomic mass may be considered to be the total mass of protons, neutrons and electrons in a single atom (when the atom is motionless.) The Atomic mass is sometimes incorrectly used as a synonym of relative Atomic mass, average Atomic mass and atomic weight; however, these differ subtly from the Atomic mass.

Chapter 22. The Transition Elements and Coordination Compounds

Chapter 22. The Transition Elements and Coordination Compounds

Complex
In chemistry, a Complex is a structure consisting of a central atom or molecule connected to surrounding atoms or molecules. Originally, a Complex implied a reversible association of molecules, atoms, or ions through weak chemical bonds. As applied to coordination chemistry, this meaning has evolved.

Covalent radius
The Covalent radius, r_{cov}, is a measure of the size of atom that forms part of a covalent bond. It is measured either in picometres (pm) or ångströms (Å), with 1 Å = 100 pm.
In principle, the sum of the two covalent radii should equal the covalent bond length between two atoms.

Principle
A Principle is a comprehensive and fundamental law, doctrine, or assumption. It can be a rule or code of conduct. It can be a law or fact of nature underlying the working of an artificial device.

EDTA
EDTA is a widely used acronym for the chemical compound ethylenediaminetetraacetic acid EDTA is a polyamino carboxylic acid with the formula $[CH_2N(CH_2CO_2H)_2]_2$. This colourless, water-soluble solid is produced on a large scale for many applications.

Molecular orbital
In chemistry, a Molecular orbital is a mathematical function that describes the wave-like behavior of an electron in a molecule. This function can be used to calculate chemical and physical properties such as the probability of finding an electron in any specific region. The use of the term "orbital" was first used in English by Robert S. Mulliken in 1925 as the English translation of Schrödinger"s use of the German word, "Eigenfunktion".

Nuclear reaction
In nuclear physics, a Nuclear reaction is the process in which two nuclei or nuclear particles collide to produce products different from the initial particles. In principle a reaction can involve more than three particles colliding, but because the probability of three or more nuclei to meet at the same time at the same place is much less than for two nuclei, such an event is exceptionally rare. While the transformation is spontaneous in the case of radioactive decay, it is initiated by a particle in the case of a Nuclear reaction.

Tungsten
Tungsten is a chemical element with the chemical symbol W and atomic number 74.
A steel-gray metal, Tungsten is found in several ores, including wolframite and scheelite. It is remarkable for its robust physical properties, especially the fact that it has the highest melting point of all the non-alloyed metals and the second highest of all the elements after carbon.

Cerium
Cerium is a chemical element with the symbol Ce and atomic number 58. It is a soft, silvery, ductile metal which easily oxidizes in air. Cerium was named after the dwarf planet Ceres.

Hafnium
Hafnium is a chemical element with the symbol Hf and atomic number 72. A lustrous, silvery gray, tetravalent transition metal, Hafnium chemically resembles zirconium and is found in zirconium minerals. Its existence was predicted by Dmitri Mendeleev in 1869.

Chapter 22. The Transition Elements and Coordination Compounds

Chapter 22. The Transition Elements and Coordination Compounds

Ionization	Ionization is the physical process of converting an atom or molecule into an ion by adding or removing charged particles such as electrons or other ions. This is often confused with dissociation (chemistry.)
	The process works slightly differently depending on whether an ion with a positive or a negative electric charge is being produced.
Lanthanide contraction	Lanthanide contraction is a term used in chemistry to describe different but closely related concepts associated with smaller than expected ionic radii of the elements in the lanthanide series (atomic number 58, Cerium to 71, Lutetium.)
	Very simply, the effect results from poor shielding of nuclear charge by 4f electrons.
	In single-electron atoms, the average separation of an electron from the nucleus is determined by the subshell it belongs to, and decreases with increasing charge on the nucleus; this in turn leads to a decrease in atomic radius.
Lutetium	Lutetium is a chemical element with the symbol Lu and atomic number 71. A silvery-white rare metal, Lutetium has the smallest atomic radius and is the heaviest and hardest member of the rare-earth group. One of its radioactive isotopes is used in nuclear technology to determine the age of meteorites.
Yttrium	Yttrium is a chemical element with symbol Y and atomic number 39. It is a silvery-metallic transition metal chemically similar to the lanthanoids and has historically been classified as a rare earth element. Yttrium is almost always found combined with the lanthanoids in rare earth minerals and is never found in nature as a free element.
Manganese	Manganese is a chemical element, designated by the symbol Mn. It has the atomic number 25. It is found as a free element in nature , and in many minerals.
Chemistry	In the history of science, the etymology of the word Chemistry is a debatable issue. It is agreed that the word "alchemy" is a European one, derived from Arabic, but the origin of the root word, chem, is uncertain. Words similar to it have been found in most ancient languages, with different meanings, but conceivably somehow related to alchemy.
Chlorous acid	Chlorous acid is a chemical compound with the formula $HClO_2$. It is a weak acid. Chlorine possesses oxidation state +3 in this acid.
Hydrochloric acid	Hydrochloric acid is the solution of hydrogen chloride (HCl) in water. It is a highly corrosive, strong mineral acid and has major industrial uses. It is found naturally in gastric acid.
Potassium	Potassium is a chemical element. It has the symbol K , atomic number 19, and atomic mass 39.0983. Potassium was first isolated from potash.

Chapter 22. The Transition Elements and Coordination Compounds

Chapter 22. The Transition Elements and Coordination Compounds

Sodium	Sodium is a metallic element with a symbol Na and atomic number 11. It is a soft, silvery-white, highly reactive metal and is a member of the alkali metals within "group 1" (formerly known as "group IA".) It has only one stable isotope, ^{23}Na.
Acid	An Acid is traditionally considered any chemical compound that, when dissolved in water, gives a solution with a hydrogen ion activity greater than in pure water, i.e. a pH less than 7.0. That approximates the modern definition of Johannes Nicolaus Brønsted and Martin Lowry, who independently defined an Acid as a compound which donates a hydrogen ion (H^+) to another compound (called a base.) Common examples include acetic Acid and sulfuric Acid (used in car batteries.)
Barium	Barium is a chemical element. It has the symbol Ba, and atomic number 56. Barium is a soft silvery metallic alkaline earth metal.
Ion	An Ion is an atom or molecule where the total number of electrons is not equal to the total number of protons, giving it a net positive or negative electrical charge. Since protons are positively charged and electrons are negatively charged, if there are more electrons than protons, the atom or molecule will be negatively charged. This is called an an Ion , from the Greek á¼€vÎ¬ , meaning "up".
Precipitation	Precipitation is the formation of a solid in a solution during a chemical reaction. When the reaction occurs, the solid formed is called the precipitate, and the liquid remaining above the solid is called the supernate. Natural methods of Precipitation include settling or sedimentation, where a solid forms over a period of time due to ambient forces like gravity or centrifugation.
Chromic acid	Chromic acid generally refers to a collection of compounds generated by the acidification of solutions containing chromate and dichromate anions or the dissolving of chromium trioxide in sulfuric acid. Often the species are assigned the formulas H_2CrO_4 and $H_2Cr_2O_7$. The anhydride of these " Chromic acid s" is chromium trioxide, also called chromium(VI) oxide; industrially, this compound is sometimes sold as Chromic acid " Regardless of its exact formula, Chromic acid features chromium in an oxidation state of +6 (or VI), often referred to as hexavalent chromium.
Nitric acid	Nitric acid, also known as aqua fortis and spirit of nitre, is a highly corrosive and toxic strong acid that can cause severe burns. Colorless when pure, older samples tend to acquire a yellow cast due to the accumulation of oxides of nitrogen. If the solution contains more than 86% Nitric acid, it is referred to as fuming Nitric acid.
Freezing	In physical science, Freezing or solidification is the process in which a liquid turns into a solid when cold enough. The Freezing point is the temperature at which this happens. Melting, the process of turning a solid to a liquid, is almost the exact opposite of Freezing.

Chapter 22. The Transition Elements and Coordination Compounds

Chapter 22. The Transition Elements and Coordination Compounds

Sulfuric acid	Sulfuric (or sulphuric) acid, H_2SO_4, is a strong mineral acid. It is soluble in water at all concentrations. Sulfuric acid has many applications, and is one of the top products of the chemical industry.
Lewis acid	A Lewis acid is a chemical compound, A, that can accept a pair of electrons from a Lewis base, B, that acts as an electron-pair donor, forming an adduct, AB. $A + :B \rightarrow A\text{--}B$ Gilbert N. Lewis proposed this definition, which is based on chemical bonding theory, in 1923. Brønsted-Lowry acid-base theory was published in the same year. The two theories are distinct but complementary to each other as a Lewis base is also a Brønsted-Lowry base, but a Lewis acid need not be a Brønsted-Lowry acid.
Base	In chemistry, a Base is most commonly thought of as an aqueous substance that can accept hydrogen ions. A Base is also often referred to as an alkali if OH^- ions are involved. This refers to the Brønsted-Lowry theory of acids and bases.
Coordination isomerism	Coordination isomerism is a form of structural isomerism in which the composition of the complex ion varies.
Coordination number	The Coordination number of an atom in a molecule or a crystal is the integer number of its nearest neighbours. This number is determined somewhat differently for molecules and for crystals. In chemistry the emphasis is on molecules (or ions) with defined bonding structures, and the Coordination number of an atom is determined by simply counting the other atoms to which it is bonded (by either single or multiple bonds.)
Ligand	In chemistry, a Ligand is either an atom, ion, or molecule that binds to a central metal to produce a coordination complex. The bonding between the metal and Ligand generally involves formal donation of one or more of the Ligand"s electrons. The metal-Ligand bonding ranges from covalent to more ionic.
Monodentate	A Monodentate ligand is a ligand which forms only one bond with the central atom, usually a metal ion. A Monodentate ligand is also sometimes called a "unidentate ligand" from the root words meaning "one tooth". A Monodentate ligand, like other types of ligands, can be a neutral molecule or an ion with a lone pair.
Ethylenediamine	Ethylenediamine is the organic compound with the formula $C_2H_4(NH_2)_2$. This colorless liquid with an ammonia-like odor is a strongly basic amine. It is a widely used building block in chemical synthesis, with approximately 500,000,000 kg being produced in 1998.

Chapter 22. The Transition Elements and Coordination Compounds

Chapter 22. The Transition Elements and Coordination Compounds

Platinum	Platinum is a chemical element with the chemical symbol Pt and an atomic number of 78." It is in Group 10 of the periodic table of elements. A dense, malleable, ductile, precious, gray-white transition metal, Platinum is resistant to corrosion and occurs in some nickel and copper ores along with some native deposits. Platinum is used in jewelry, laboratory equipment, electrical contacts and electrodes, Platinum resistance thermometers, dentistry equipment, and catalytic converters.
Salt	Salt is a dietary mineral composed primarily of sodium chloride that is essential for animal life, but toxic to most land plants. Salt flavor is one of the basic tastes, an important preservative and a popular food seasoning.
Cobalt	Cobalt is a hard, lustrous, grey metal, a chemical element with symbol Co and atomic number 27. Although Cobalt based colors and pigments have been used since ancient times for making jewelry and paints, and miners have long used the name kobold ore for some minerals, the free metallic Cobalt was not prepared and discovered until 1735 by Georg Brandt.
Square planar	The square planar molecular geometry in chemistry describes the stereochemistry (spatial arrangement of atoms) that is adopted by certain chemical compounds. As the name suggests, molecules of this geometry have their atoms positioned at the corners of a square on the same plane about a central atom. The addition of two ligands to linear compounds, ML_2, can afford square planar complexes.
Catalyst	Catalyst is a science centre and museum devoted to the chemical industry. Its full title is Catalyst Science Discovery Centre. It is located in Widnes, Cheshire, in the north west of England, and situated on the north bank of the River Mersey (grid reference SJ512841.)
Enantiomer	In chemistry, an Enantiomer is one of two stereoisomers that are non-superposable complete mirror images of each other, much as one"s left and right hands are "the same" but opposite. Enantiopure compounds refer to a sample having within the limits of detection, molecules of only one chirality. Enantiomer s have, when present in a symmetric environment, identical chemical and physical properties except for their ability to rotate plane-polarized light (+/-) by equal amounts but in opposite directions.
Racemic mixture	In chemistry, a Racemic mixture is one that has equal amounts of left- and right-handed enantiomers of a chiral molecule. The first known Racemic mixture was "racemic acid," which Louis Pasteur found to be a mixture of the two enantiomeric isomers of tartaric acid. A racemate is optically inactive, meaning that it does not rotate plane-polarized light.
Emission	In physics, Emission is the process by which the energy of a photon is released by another entity, for example, by an atom whose electrons make a transition between two electronic energy levels. The emitted energy is in the form of a photon.

Chapter 22. The Transition Elements and Coordination Compounds

Chapter 22. The Transition Elements and Coordination Compounds

	The emittance of an object quantifies how much light is emitted by it.
Covalent bond	A Covalent bond is a form of chemical bonding that is characterized by the sharing of pairs of electrons between atoms, or between atoms and other Covalent bond s. In short, attraction-to-repulsion stability that forms between atoms when they share electrons is known as Covalent bond ing. Covalent bond ing includes many kinds of interaction, including σ-bonding, π-bonding, metal to non-metal bonding, agostic interactions, and three-center two-electron bonds.
Valence	In chemistry, Valence is a measure of the number of chemical bonds formed by the atoms of a given element. Over the last century, the concept of Valence evolved into a range of approaches for describing the chemical bond, including Lewis structures, Valence bond theory, molecular orbitals, Valence shell electron pair repulsion theory and all the advanced methods of quantum chemistry. The etymology of the word "Valence" is from 1425, meaning "extract, preparation," from Latin valentia "strength, capacity," and the chemical meaning referring to the "combining power of an element" is recorded from 1884, from German Valenz.
Valence bond theory	In chemistry, Valence bond theory is one of two basic theories, along with molecular orbital theory, that developed to use the methods of quantum mechanics to explain chemical bonding. It focuses on how the atomic orbitals of the dissociated atoms combine on molecular formation to give individual chemical bonds. In contrast, molecular orbital theory has orbitals that cover the whole molecule.
Crystal	A Crystal or Crystal line solid is a solid material whose constituent atoms, molecules, or ions are arranged in an orderly repeating pattern extending in all three spatial dimensions. The scientific study of Crystal s and Crystal formation is Crystal lography. The process of Crystal formation via mechanisms of Crystal growth is called Crystal lization or solidification.
Crystal field theory	Crystal field theory is a model that describes the electronic structure of transition metal compounds, all of which can be considered coordination complexes. Crystal field theory successfully accounts for some magnetic properties, colours, hydration enthalpies, and spinel structures of transition metal complexes, but it does not attempt to describe bonding. Crystal field theory was developed by physicists Hans Bethe and John Hasbrouck van Vleck in the 1930s.
Ligand field theory	Ligand field theory describes the bonding, orbital arrangement, and other characteristics of coordination complexes. It represents an application of molecular orbital theory to transition metal complexes. A transition metal ion has nine valence atomic orbitals, five nd, one (n+1)s, and three (n+1)p orbitals.

Chapter 22. The Transition Elements and Coordination Compounds

Chapter 22. The Transition Elements and Coordination Compounds

Molecular orbital theory

In chemistry, Molecular orbital theory is a method for determining molecular structure in which electrons are not assigned to individual bonds between atoms, but are treated as moving under the influence of the nuclei in the whole molecule. In this theory, each molecule has a set of molecular orbitals, in which it is assumed that the molecular orbital wave function ψ_f may be written as a simple weighted sum of the n constituent atomic orbitals χ_i, according to the following equation:

$$\psi_j = \sum_{i=1}^{n} c_{ij} \chi_i$$

The c_{ij} coefficients may be determined numerically by substitution of this equation into the Schrödinger equation and application of the variational principle. This method is called the linear combination of atomic orbitals approximation and is used in computational chemistry.

Hydrocarbon

In organic chemistry, a Hydrocarbon is an organic compound consisting entirely of hydrogen and carbon. With relation to chemical terminology, aromatic Hydrocarbon s or arenes, alkanes, alkenes and alkyne-based compounds composed entirely of carbon or hydrogen are referred to as "pure" Hydrocarbon s, whereas other Hydrocarbon s with bonded compounds or impurities of sulfur or nitrogen, are referred to as "impure", and remain somewhat erroneously referred to as Hydrocarbon s.

Hydrocarbon s are referred to as consisting of a "backbone" or "skeleton" composed entirely of carbon and hydrogen and other bonded compounds, and have a functional group that generally facilitates combustion.

Spectrochemical series

A Spectrochemical series is a list of ligands ordered on ligand strength and a list of metal ions based on oxidation number, group and its identity. In crystal field theory, ligands modify the difference in energy between the d orbitals (Δ - ligand-field splitting parameter, or crystal-field splitting parameter), what is mainly reflected in differences in color of similar metal-ligand complexes.

A partial Spectrochemical series listing of ligands from small Δ to large Δ:

$I^- < Br^- < S^{2-} < SCN^- < Cl^- < NO_3^- < N_3^- < F^- < OH^- < C_2O_4^{2-} \approx H_2O < NCS^- < CH_3CN <$ py (pyridine) $< NH_3 <$ en (ethylenediamine) $<$ bipy (2,2"-bipyridine) $<$ phen (1,10-phenanthroline) $< NO_2^- < PPh_3 < CN^- \approx CO$

Oxygen

Oxygen and -γενÎ®ς (-genÄ"s) (producer, literally begetter) is the element with atomic number 8 and represented by the symbol O. It is a member of the chalcogen group on the periodic table, and is a highly reactive nonmetallic period 2 element that readily forms compounds (notably oxides) with almost all other elements. At standard temperature and pressure two atoms of the element bind to form di Oxygen , a colorless, odorless, tasteless diatomic gas with the formula O_2. Oxygen is the third most abundant element in the universe by mass after hydrogen and helium and the most abundant element by mass in the Earth"s crust.

Chapter 22. The Transition Elements and Coordination Compounds

Chapter 23. Organic Chemistry

Carbon

Carbon is the chemical element with symbol C and atomic number 6. As a member of group 14 on the periodic table, it is nonmetallic and tetravalent--making four electrons available to form covalent chemical bonds. There are three naturally occurring isotopes, with ^{12}C and ^{13}C being stable, while ^{14}C is radioactive, decaying with a half-life of about 5730 years.

Organic chemistry

Organic chemistry is a discipline within chemistry which involves the scientific study of the structure, properties, composition, reactions, and preparation (by synthesis or by other means) of chemical compounds that contain carbon. These compounds may contain any number of other elements, including hydrogen, nitrogen, oxygen, the halogens as well as phosphorus, silicon and sulphur.

The original definition of "organic" chemistry came from the misconception that organic compounds were always related to life processes.

Acid

An Acid is traditionally considered any chemical compound that, when dissolved in water, gives a solution with a hydrogen ion activity greater than in pure water, i.e. a pH less than 7.0. That approximates the modern definition of Johannes Nicolaus Brønsted and Martin Lowry, who independently defined an Acid as a compound which donates a hydrogen ion (H^+) to another compound (called a base.) Common examples include acetic Acid and sulfuric Acid (used in car batteries.)

Chemistry

In the history of science, the etymology of the word Chemistry is a debatable issue. It is agreed that the word "alchemy" is a European one, derived from Arabic, but the origin of the root word, chem, is uncertain. Words similar to it have been found in most ancient languages, with different meanings, but conceivably somehow related to alchemy.

Product

A Product is a substance that forms as a result of a biological- or chemical reaction. While the end Product of some chemical reactions may be the result of a relatively rapid reaction, nanoseconds to seconds, chemical equilibria in complex systems may require years or even centuries to be established. For example, equilibria in groundwater systems with multiple components are achieved on timescales of millennia, if ever.

Group

In chemistry, a Group is a vertical column in the periodic table of the chemical elements. The name family is derived from the fact that the elements share similar characteristics and traits, just as members of any human family would. There are 18 groups in the standard periodic table.

Hydrocarbon

In organic chemistry, a Hydrocarbon is an organic compound consisting entirely of hydrogen and carbon. With relation to chemical terminology, aromatic Hydrocarbon s or arenes, alkanes, alkenes and alkyne-based compounds composed entirely of carbon or hydrogen are referred to as "pure" Hydrocarbon s, whereas other Hydrocarbon s with bonded compounds or impurities of sulfur or nitrogen, are referred to as "impure", and remain somewhat erroneously referred to as Hydrocarbon s.

Hydrocarbon s are referred to as consisting of a "backbone" or "skeleton" composed entirely of carbon and hydrogen and other bonded compounds, and have a functional group that generally facilitates combustion.

Chapter 23. Organic Chemistry

Chapter 23. Organic Chemistry

Molecular geometry	Molecular geometry or molecular structure is the three-dimensional arrangement of the atoms that constitute a molecule. It determines several properties of a substance including its reactivity, polarity, phase of matter, color, magnetism, and biological activity. The Molecular geometry can be determined by various spectroscopic methods and diffraction methods.
Methanation	Methanation is a physical-chemical process to generate Methane from a mixture of various gases out of biomass fermentation or thermo-chemical gasification. The main components are carbon monoxide and hydrogen. The following main process describes the Methanation: $$CO + 3\,H_2 \rightarrow CH_4 + H_2O$$ This process is used for the generation of biogenous natural gas substitute, which can be fed into the gas grid. Methanation is the reverse reaction of steam methane reforming, which converts methane into synthesis gas.
Boiling	Boiling, a type of phase transition, is the rapid vaporization of a liquid, which typically occurs when a liquid is heated to its Boiling point, the temperature at which the vapor pressure of the liquid is equal to the pressure exerted on the liquid by the surrounding environmental pressure. Thus, a liquid may also boil when the pressure of the surrounding atmosphere is sufficiently reduced, such as the use of a vacuum pump or at high altitudes. Boiling occurs in three characteristic stages, which are nucleate, transition and film Boiling.
Melting	Melting is a physical process that results in the phase change of a substance from a solid to a liquid. The internal energy of a solid substance is increased, typically by the application of heat or pressure, resulting in a rise of its temperature to the Melting point, at which the rigid ordering of molecular entities in the solid breaks down to a less-ordered state and the solid liquefies. An object that has melted completely is molten.
Melting point	The Melting point of a solid is the temperature range at which it changes state from solid to liquid. At the Melting point the solid and liquid phase exist in equilibrium. When considered as the temperature of the reverse change from liquid to solid, it is referred to as the freezing point.
Solution	In chemistry, a Solution is a homogeneous mixture composed of two or more substances. In such a mixture, a solute is dissolved in another substance, known as a solvent. Gases may dissolve in liquids, for example, carbon dioxide or oxygen in water.
Intermolecular forces	In physics, chemistry, and biology, Intermolecular forces are forces that act between stable molecules or between functional groups of macromolecules. Intermolecular forces include momentary attractions between molecules, diatomic free elements, and individual atoms. They differ from covalent and ionic bonding in that they are not stable, but are caused by momentary polarization of particles.

Chapter 23. Organic Chemistry

Chapter 23. Organic Chemistry

Isomer
: In chemistry, Isomer s are compounds with the same molecular formula but different structural formula. Isomer s do not necessarily share similar properties unless they also have the same functional groups. This should not be confused with a nuclear Isomer which involves a nucleus at different states of excitement.

Source
: Source Holdings Ltd. is a specialist European-based provider of Exchange Traded Products (ETPs), including Exchange Traded Funds (ETFs) and Exchange Traded Commodities (ETCs.) The company was founded in 2008 by Bank of America Merrill Lynch, Goldman Sachs and Morgan Stanley and is headquartered inside the City of London, on 88 Wood Street.
Source launched its first product offering on the 20th April 2009, although the 3 founders are reported to have started working on the project in early 2008 .

Distillation
: Distillation is a method of separating mixtures based on differences in their volatilities in a boiling liquid mixture. Distillation is a unit operation, or a physical separation process, and not a chemical reaction.
Commercially, Distillation has a number of uses.

Fractional distillation
: Fractional distillation is the separation of a mixture into its component parts such as in separating chemical compounds by their boiling point by heating them to a temperature at which several fractions of the compound will evaporate. It is a special type of distillation. Generally the component parts boil at less than 25 °C from each other under a pressure of one atmosphere (atm.)

Cracking
: In petroleum geology and chemistry, Cracking is the process whereby complex organic molecules such as kerogens or heavy hydrocarbons are broken down into simpler molecules (e.g. light hydrocarbons) by the breaking of carbon-carbon bonds in the precursors. The rate of Cracking and the end products are strongly dependent on the temperature and presence of any catalysts. Cracking, also referred to as pyrolysis, is the breakdown of a large alkane into smaller, more useful alkanes and an alkene.

Hydrogen
: Hydrogen is the chemical element with atomic number 1. It is represented by the symbol H. At standard temperature and pressure, Hydrogen is a colorless, odorless, nonmetallic, tasteless, highly flammable diatomic gas with the molecular formula H_2. With an atomic weight of 1.007 94 u, Hydrogen is the lightest element.

Oxygen
: Oxygen and -γενÎ®ς (-genÄ"s) (producer, literally begetter) is the element with atomic number 8 and represented by the symbol O. It is a member of the chalcogen group on the periodic table, and is a highly reactive nonmetallic period 2 element that readily forms compounds (notably oxides) with almost all other elements. At standard temperature and pressure two atoms of the element bind to form di Oxygen , a colorless, odorless, tasteless diatomic gas with the formula O_2. Oxygen is the third most abundant element in the universe by mass after hydrogen and helium and the most abundant element by mass in the Earth"s crust.

Chapter 23. Organic Chemistry

Chapter 23. Organic Chemistry

Transuranium elements	In chemistry, Transuranium elements are the chemical elements with atomic numbers greater than 92 None of these elements are stable; they decay radioactively into other elements. Of the elements with atomic numbers 1 to 92, all but four occur in easily detectable quantities on earth, having stable, or very long half life isotopes, or are created as common products of the decay of uranium.
Hydrogenation	Hydrogenation is the chemical reaction that results from the addition of hydrogen (H_2.) The process is usually employed to reduce or saturate organic compounds. The process typically constitutes the addition of pairs of hydrogen atoms to a molecule.
Polyethylene	Polyethylene or polythene (IUPAC name polyethene or poly(methylene)) is a thermoplastic commodity heavily used in consumer products (notably the plastic shopping bag.) Over 60 million tons of the material are produced worldwide every year. Polyethylene is a polymer consisting of long chains of the monomer ethylene (IUPAC name ethene.)
Bromine	Bromine , Greek: βρά¿¶μος, brómos, meaning "stench "), is a chemical element with the symbol Br and atomic number 35. A halogen element, Bromine is a reddish-brown volatile liquid at standard room temperature that is intermediate in reactivity between chlorine and iodine. Bromine vapours are corrosive and toxic.
Potassium	Potassium is a chemical element. It has the symbol K , atomic number 19, and atomic mass 39.0983. Potassium was first isolated from potash.
Benzene	Benzene, or benzol, is an organic chemical compound and a known carcinogen with the molecular formula C_6H_6. It is sometimes abbreviated Ph-H. Benzene is a colorless and highly flammable liquid with a sweet smell and a relatively high melting point. Because it is a known carcinogen, its use as an additive in gasoline is now limited, but it is an important industrial solvent and precursor in the production of drugs, plastics, synthetic rubber, and dyes.
Calcium	Calcium is the chemical element with the symbol Ca and atomic number 20. It has an atomic mass of 40.078 amu. Calcium is a soft grey alkaline earth metal, and is the fifth most abundant element by mass in the Earth"s crust.
Molecular orbital	In chemistry, a Molecular orbital is a mathematical function that describes the wave-like behavior of an electron in a molecule. This function can be used to calculate chemical and physical properties such as the probability of finding an electron in any specific region. The use of the term "orbital" was first used in English by Robert S. Mulliken in 1925 as the English translation of Schrödinger"s use of the German word, "Eigenfunktion".

Chapter 23. Organic Chemistry

Chapter 23. Organic Chemistry

Carbide	In chemistry, a Carbide is a compound composed of carbon and a less electronegative element. Carbide s can be generally classified by chemical bonding type as follows: (i) salt-like, (ii) covalent compounds, (iii) interstitial compounds, and (iv) "intermediate" transition metal Carbide s. Examples include calcium Carbide silicon Carbide tungsten Carbide (often called simply Carbide , and cementite, each used in key industrial applications.
Cinnamaldehyde	Cinnamic aldehyde or Cinnamaldehyde is the chemical compound that gives cinnamon its flavor and odor. Cinnamaldehyde occurs naturally in the bark of cinnamon trees and other species of the genus Cinnamomum like camphor and cassia. These trees are the natural source of cinnamon, and the essential oil of cinnamon bark is about 90% Cinnamaldehyde.
Polycyclic aromatic hydrocarbons	Polycyclic aromatic hydrocarbons (Polycyclic aromatic hydrocarbons s) are chemical compounds that consist of fused aromatic rings and do not contain heteroatoms or carry substituents. Polycyclic aromatic hydrocarbons s occur in oil, coal, and tar deposits, and are produced as byproducts of fuel burning (whether fossil fuel or biomass.) As a pollutant, they are of concern because some compounds have been identified as carcinogenic, mutagenic, and teratogenic.
Molecular model	A Molecular model in this article, is a physical model that represents molecules and their processes. The creation of mathematical models of molecular properties and behaviour is Molecular model ling, and their graphical depiction is molecular graphics, but these topics are closely linked and each uses techniques from the others. In this article, Molecular model will primarily refer to systems containing more than one atom and where nuclear structure is neglected.
Ion	An Ion is an atom or molecule where the total number of electrons is not equal to the total number of protons, giving it a net positive or negative electrical charge. Since protons are positively charged and electrons are negatively charged, if there are more electrons than protons, the atom or molecule will be negatively charged. This is called an an Ion , from the Greek á¼€vÎ¬ , meaning "up".
Reactivity	Reactivity refers to the rate at which a chemical substance tends to undergo a chemical reaction in time. In pure compounds, Reactivity is regulated by the physical properties of the sample. For instance, grinding a sample to a higher specific surface area increases its Reactivity.
Ethanol	Ethanol pure alcohol, grain alcohol is a volatile, flammable, colorless liquid. It is a psychoactive drug, best known as the type of alcohol found in alcoholic beverages and in modern thermometers. Ethanol is one of the oldest recreational drugs.

Chapter 23. Organic Chemistry

Chapter 23. Organic Chemistry

Mole	The mole is a unit of amount of substance: it is an SI base unit, and one of the few units used to measure this physical quantity. The name "mole" was coined in German by Wilhelm Ostwald in 1893, although the related concept of equivalent mass had been in use at least a century earlier. The name is assumed to be derived from the word Molekül (molecule.)
Nitrogen	Nitrogen is a chemical element that has the symbol N and atomic number 7 and atomic mass 14.00674 u. Elemental Nitrogen is a colorless, odorless, tasteless and mostly inert diatomic gas at standard conditions, constituting 78% by volume of Earth"s atmosphere. Many industrially important compounds, such as ammonia, nitric acid, organic nitrates , and cyanides, contain Nitrogen.
Carboxylic acids	Carboxylic acids are organic acids characterized by the presence of a carboxyl group, which has the formula -C(=O)OH, usually written -COOH or $-CO_2H$. Carboxylic acids are Brønsted-Lowry acids -- they are proton donors. Salts and anions of Carboxylic acids are called carboxylates. The simplest series of Carboxylic acids are the alkanoic acids, R-COOH, where R is a hydrogen or an alkyl group.
Ammonia	Ammonia is a compound of nitrogen and hydrogen with the formula NH_3. It is normally encountered as a gas with a characteristic pungent odor. Ammonia contributes significantly to the nutritional needs of terrestrial organisms by serving as a precursor to foodstuffs and fertilizers.
Aqueous solution	An Aqueous solution is a solution in which the solvent is water. It is usually shown in chemical equations by appending (aq) to the relevant formula. The word aqueous means pertaining to, related to, similar to, or dissolved in water.
Base	In chemistry, a Base is most commonly thought of as an aqueous substance that can accept hydrogen ions. A Base is also often referred to as an alkali if OH^- ions are involved. This refers to the Brønsted-Lowry theory of acids and bases.
Ionization	Ionization is the physical process of converting an atom or molecule into an ion by adding or removing charged particles such as electrons or other ions. This is often confused with dissociation (chemistry.) The process works slightly differently depending on whether an ion with a positive or a negative electric charge is being produced.

Chapter 24. Polymer Materials: Synthetic and Biological

Polyethylene	Polyethylene or polythene (IUPAC name polyethene or poly(methylene)) is a thermoplastic commodity heavily used in consumer products (notably the plastic shopping bag.) Over 60 million tons of the material are produced worldwide every year. Polyethylene is a polymer consisting of long chains of the monomer ethylene (IUPAC name ethene.)
Polymer	A Polymer is a large molecule composed of repeating structural units typically connected by covalent chemical bonds. While Polymer in popular usage suggests plastic, the term actually refers to a large class of natural and synthetic materials with a variety of properties. Due to the extraordinary range of properties accessible in Polymer ic materials , they have come to play an essential and ubiquitous role in everyday life - from plastics and elastomers on the one hand to natural bio Polymer s such as DNA and proteins that are essential for life on the other.
Activity	In chemical thermodynamics Activity is a measure of the "effective concentration" of a species in a mixture. By convention, it is a dimensionless quantity. The Activity of pure substances in condensed phases (solid or liquids) is normally taken as unity.
Polypropylene	Polypropylene or polypropene (PP) is a thermoplastic polymer, made by the chemical industry and used in a wide variety of applications, including packaging, textiles (e.g. ropes, thermal underwear and carpets), stationery, plastic parts and reusable containers of various types, laboratory equipment, loudspeakers, automotive components, and polymer banknotes. An addition polymer made from the monomer propylene, it is rugged and unusually resistant to many chemical solvents, bases and acids. In 2007, the global market for Polypropylene had a volume of 45.1 million tons which led to a turnover of about 65 billion US $ (47,4 billion â,¬.)
Product	A Product is a substance that forms as a result of a biological- or chemical reaction. While the end Product of some chemical reactions may be the result of a relatively rapid reaction, nanoseconds to seconds, chemical equilibria in complex systems may require years or even centuries to be established. For example, equilibria in groundwater systems with multiple components are achieved on timescales of millennia, if ever.
Isomer	In chemistry, Isomer s are compounds with the same molecular formula but different structural formula. Isomer s do not necessarily share similar properties unless they also have the same functional groups. This should not be confused with a nuclear Isomer which involves a nucleus at different states of excitement.

Chapter 24. Polymer Materials: Synthetic and Biological

Chapter 24. Polymer Materials: Synthetic and Biological

Latex	Latex is a name collectively given to a group of similar preparations consisting of stable dispersions of polymer microparticles in a liquid matrix (usually water.) Latexes may be natural or synthetic. Synthetic latexes are usually produced by emulsion polymerization using a variety of initiators, surfactants and monomers; the latter commonly include: • vinyl acetate • SBR (styrene-butadiene) • acrylates while more exotic formulations include allylic compounds, vinyl malate or VEOVAs. Latexes are used in coatings (e.g. Latex paint) and glues because they solidify by coalescence of the polymer particles as the water evaporates, and therefore can form films without releasing potentially toxic organic solvents in the environment.
Condensation	Condensation is the change of the physical state of aggregation (or simply state) of matter from gaseous phase into liquid phase. When the transition happens from the gaseous phase into the solid phase directly, bypassing the liquid phase, the change is called deposition. Condensation commonly occurs when a vapor is cooled to its dew point, but the dew point can also be reached through compression.
Phase transition	In thermodynamics, a Phase transition is the transformation of a thermodynamic system from one phase to another. At a Phase transition point, physical properties may undergo abrupt change: for instance, the volume of the two phases may be vastly different as is illustrated by the boiling of liquid water to form steam. The term is most commonly used to describe transitions between solid, liquid and gaseous states of matter, in rare cases including plasma.
Acid	An Acid is traditionally considered any chemical compound that, when dissolved in water, gives a solution with a hydrogen ion activity greater than in pure water, i.e. a pH less than 7.0. That approximates the modern definition of Johannes Nicolaus Brønsted and Martin Lowry, who independently defined an Acid as a compound which donates a hydrogen ion (H^+) to another compound (called a base.) Common examples include acetic Acid and sulfuric Acid (used in car batteries.)
Alan Jay Heeger	Alan Jay Heeger is an American physicist, academic and Nobel Prize laureate in chemistry. Heeger was born in Sioux City, Iowa. He earned a B.S. in physics and mathematics from the University of Nebraska in 1957, and a Ph.D in physics from the University of California, Berkeley in 1961.

Chapter 24. Polymer Materials: Synthetic and Biological

Chapter 24. Polymer Materials: Synthetic and Biological

Polyacetylene	Polyacetylene (IUPAC name: polyethyne) is an organic polymer with the repeat unit $(C_2H_2)_n$. The high electrical conductivity discovered for these polymers in the 1970"s accelerated interest in the use of organic compounds in microelectronics (organic electronics.) Polyacetylene s are also known where the H atoms are replaced with alkyl groups.
Graphite	Graphite " href="/wiki/Cleavage_(crystal)">loose interlamellar coupling between sheets in the structure, in fact in a vacuum environment (such as in technologies for use in space), Graphite was found to be a very poor lubricant. This fact led to the discovery that Graphite"s lubricity is due to adsorbed air and water between the layers, unlike other layered dry lubricants such as molybdenum disulfide. Recent studies suggest that an effect called superlubricity can also account for this effect.
Gallium	Gallium is a chemical element that has the symbol Ga and atomic number 31. Elemental Gallium does not occur in nature, but as the Ga salt, in trace amounts in bauxite and zinc ores. A soft silvery metallic poor metal, elemental Gallium is a brittle solid at low temperatures.
Gallium arsenide	Gallium arsenide is a compound of two elements, gallium and arsenic. It is an important semiconductor and is used to make devices such as microwave frequency integrated circuits (ie, MMICs), infrared light-emitting diodes, laser diodes and solar cells. Gallium arsenide can be prepared from the elements and a number of industrial processes use this, for example: - the crystal growth using a horizontal zone furnace (Bridgman-Stockbarger technique) where Ga and Arsenic vapor react and deposit on a seed crystal at the cooler end of the furnace. - LEC (liquid encapsulated Czochralski) growth Alternative methods for producing films of GaAs include: - VPE reaction of gaseous gallium metal and arsenic trichloride $2Ga + 2AsCl_3 \rightarrow 2GaAs + 3Cl_2$ - MOCVD reaction of trimethylgallium and arsine: $Ga(CH_3)_3 + AsH_3 \rightarrow GaAs + 3CH_4$ Wet etching of GaAs industrially uses an oxidizing agent e.g. hydrogen peroxide or bromine water, and the same strategy has been described in a patent relating to processing scrap components containing GaAs where the Ga^{3+} is complexed with a hydroxamic acid, "HA" e.g.: $GaAs + H_2O_2 + "HA" \rightarrow "GaA" \text{ complex} + H_3AsO_4 + 4H_2O$

Chapter 24. Polymer Materials: Synthetic and Biological

Chapter 24. Polymer Materials: Synthetic and Biological

	Oxidation of GaAs occurs in air and degrades performance of the semiconductor, the surface can be passivated by depositing a cubic gallium(II) sulfide layer using a tert-butyl gallium sulfide compound such as (^{t}BuGaS)$_7$.
	GaAs has some electronic properties which are superior to those of silicon.
Enantiomer	In chemistry, an Enantiomer is one of two stereoisomers that are non-superposable complete mirror images of each other, much as one"s left and right hands are "the same" but opposite. Enantiopure compounds refer to a sample having within the limits of detection, molecules of only one chirality.
	Enantiomer s have, when present in a symmetric environment, identical chemical and physical properties except for their ability to rotate plane-polarized light (+/-) by equal amounts but in opposite directions.
Peptide bond	A Peptide bond is a chemical bond formed between two molecules when the carboxyl group of one molecule reacts with the amine group of the other molecule, thereby releasing a molecule of water (H$_2$O.) This is a dehydration synthesis reaction (also known as a condensation reaction), and usually occurs between amino acids. The resulting CO-NH bond is called a Peptide bond, and the resulting molecule is an amide.
Intermolecular forces	In physics, chemistry, and biology, Intermolecular forces are forces that act between stable molecules or between functional groups of macromolecules. Intermolecular forces include momentary attractions between molecules, diatomic free elements, and individual atoms. They differ from covalent and ionic bonding in that they are not stable, but are caused by momentary polarization of particles.
Primary structure	In biochemistry, the Primary structure of a biological molecule is the exact specification of its atomic composition and the chemical bonds connecting those atoms (including stereochemistry.) For a typical unbranched, un-crosslinked biopolymer (such as a molecule of DNA, RNA or typical intracellular protein), the Primary structure is equivalent to specifying the sequence of its monomeric subunits, e.g., the nucleotide or peptide sequence. The term "Primary structure" was first coined by Linderstrøm-Lang in his 1951 Lane Medical Lectures.
Secondary structure	In biochemistry and structural biology, Secondary structure is the general three-dimensional form of local segments of biopolymers such as proteins and nucleic acids (DNA/RNA.) It does not, however, describe specific atomic positions in three-dimensional space, which are considered to be tertiary structure.
	Secondary structure is formally defined by the hydrogen bonds of the biopolymer, as observed in an atomic-resolution structure.
Tertiary structure	In biochemistry and chemistry, the Tertiary structure of a protein or any other macromolecule is its three-dimensional structure, as defined by the atomic coordinates.

Chapter 24. Polymer Materials: Synthetic and Biological

Chapter 24. Polymer Materials: Synthetic and Biological

	Tertiary structure is considered to be largely determined by the protein"s primary structure, or the sequence of amino acids of which it is composed. Efforts to predict Tertiary structure from the primary structure are known generally as protein structure prediction.
Active site	The Active site of an enzyme contains the catalytic and binding sites. The structure and chemical properties of the Active site allow the recognition and binding of the substrate. The Active site is usually a big pocket or cleft surrounded by amino acid- and other side chains at the surface of the enzyme that contains residues responsible for the substrate specificity (charge, hydrophobicity, steric hindrance) and catalytic residues which often act as proton donors or acceptors or are responsible for binding a cofactor such as PLP, TPP or NAD. The Active site is also the site of inhibition of enzymes
Enzymes	Enzymes are biomolecules that catalyze (i.e., increase the rates of) chemical reactions. Nearly all known Enzymes are proteins. However, certain RNA molecules can be effective biocatalysts too.
Enzyme catalysis	Enzyme catalysis is the catalysis of chemical reactions by specialized proteins known as enzymes. Catalysis of biochemical reactions in the cell is vital due to the very low reaction rates of the uncatalysed reactions. The mechanism of Enzyme catalysis is similar in principle to other types of chemical catalysis.
Monosaccharides	Monosaccharides are the most basic unit of carbohydrates. They are the simplest form of sugar and are usually colorless, water-soluble, crystalline solids. Some Monosaccharides have a sweet taste.
Nucleic acid	A Nucleic acid is a macromolecule composed of chains of monomeric nucleotides. In biochemistry these molecules carry genetic information or form structures within cells. The most common Nucleic acid s are deoxyribo Nucleic acid (D Nucleic acid) and ribo Nucleic acid (R Nucleic acid .)
Nucleotide	Nucleotide s are molecules that, when joined together, make up the structural units of RNA and DNA. Additionally, Nucleotide s play central roles in metabolism. In that capacity, they serve as sources of chemical energy (adenosine triphosphate and guanosine triphosphate), participate in cellular signaling (cyclic guanosine monophosphate and cyclic adenosine monophosphate), and are incorporated into important cofactors of enzymatic reactions (coenzyme A, flavin adenine di Nucleotide , flavin mono Nucleotide , and nicotinamide adenine di Nucleotide phosphate.) Figure 1: Structural elements of the most common Nucleotide s Figure 2: Ribose structure indicating numbering of carbon atoms A Nucleotide is composed of a nucleobase (nitrogenous base) and a five-carbon sugar (either ribose or 2"-deoxyribose), and one to three phosphate groups.

Chapter 24. Polymer Materials: Synthetic and Biological

Chapter 24. Polymer Materials: Synthetic and Biological

Catalysis	Catalysis is the process in which the rate of a chemical reaction is either increased or decreased by means of a chemical substance known as a catalyst. Unlike other reagents that participate in the chemical reaction, a catalyst is not consumed by the reaction itself. The catalyst may participate in multiple chemical transformations.
Iron	Iron is a chemical element with the symbol Fe and atomic number 26. Iron is a group 8 and period 4 element. Iron and Iron alloys (steels) are by far the most common metals and the most common ferromagnetic materials in everyday use.
DNA	Deoxyribonucleic acid (DNA) is a nucleic acid that contains the genetic instructions used in the development and functioning of all known living organisms and some viruses. The main role of DNA molecules is the long-term storage of information. DNA is often compared to a set of blueprints or a recipe, or a code, since it contains the instructions needed to construct other components of cells, such as proteins and RNA molecules.
Base	In chemistry, a Base is most commonly thought of as an aqueous substance that can accept hydrogen ions. A Base is also often referred to as an alkali if OH^- ions are involved. This refers to the Brønsted-Lowry theory of acids and bases.
Hydrogen	Hydrogen is the chemical element with atomic number 1. It is represented by the symbol H. At standard temperature and pressure, Hydrogen is a colorless, odorless, nonmetallic, tasteless, highly flammable diatomic gas with the molecular formula H_2. With an atomic weight of 1.007 94 u, Hydrogen is the lightest element.
Polymerase chain reaction	In molecular biology, the Polymerase chain reaction is a technique to amplify a single or few copies of a piece of DNA across several orders of magnitude, generating millions or more copies of a particular DNA sequence. The method relies on thermal cycling, consisting of cycles of repeated heating and cooling of the reaction for DNA melting and enzymatic replication of the DNA. Primers (short DNA fragments) containing sequences complementary to the target region along with a DNA polymerase (after which the method is named) are key components to enable selective and repeated amplification. As Polymerase chain reaction progresses, the DNA generated is itself used as a template for replication, setting in motion a chain reaction in which the DNA template is exponentially amplified.
Microscopy	Microscopy is the technical field of using microscopes to view samples or objects. There are three well-known branches of Microscopy, optical, electron and scanning probe Microscopy. Optical and electron Microscopy involve the diffraction, reflection, or refraction of electromagnetic radiation/electron beam interacting with the subject of study, and the subsequent collection of this scattered radiation in order to build up an image.

Chapter 24. Polymer Materials: Synthetic and Biological

CPSIA information can be obtained at www.ICGtesting.com
Printed in the USA
BVOW050528300812

299106BV00001B/177/P